For Pete

With —

May 1974.

WRECK AND RESCUE ON THE ESSEX COAST

IT WAS NOT UNTIL 1876, following the wreck of the German emigrant ship *Deutschland*, with the loss of many lives, that the Royal National Lifeboat Institution sent a lifeboat to Essex, but the story of lifesaving on this busy and most dangerous coast goes back long before that. The salvaging smacks from Harwich and the Colne had a remarkable record of life-saving among the off-lying sandbanks in the approaches to the Thames Estuary, and it was largely due to their success in this direction that lifeboats were for many years regarded as unnecessary in this area. Perhaps it was because of the smacks' good work that early attempts to form a lifeboat association in the county had no lasting success. For the past ninety years, however, the lifeboats of Essex have done extremely good work, often in difficult and hazardous conditions, and the story of their operations contains many a tale of heroism and high endeavour, and sometimes of tragedy. The story of the part Harwich played in the development of the steam lifeboat; of the rivalry between the "national" and "private" lifeboats at Walton-on-Naze; of heroism and disaster at Clacton-on-Sea; and of the Southend lifeboat's work in the approaches to the country's busiest port in peace and war has never previously been told in detail. And as well as covering the history of the salvaging smacks and of each of the county's lifeboat stations, this book tells also of the firms which built many of Britain's lifeboats, playing a most important part in the development of the modern lifeboat.

The Wreck and Rescue Series—

General Editor GRAHAME FARR

Wreck and Rescue on the Essex Coast

THE STORY OF THE ESSEX LIFEBOATS

Robert Malster

D. BRADFORD BARTON LTD
TRURO

First published 1968 by
D. BRADFORD BARTON LTD
TRURO BOOKSHOP TRURO CORNWALL

Printed by H. E. Warne Ltd St. Austell

CONTENTS

ILLUSTRATIONS

INTRODUCTION

THE *Mariners' Mirror*, the first collection of sailing directions published in England—in the year of the Armada's defeat—introduced the directions for Harwich with the words 'A perfect description and pourtraiet of the Chiefest Port of England, viz.—the river of London and of all the sandes, bankes, flattes and shoaldes unto Harwich . . .' and concluded them with 'In these coastes is very great traffique . . .'

In those days wrecks must have been frequent, but they went largely unrecorded; there were no lifesaving services and the mariners who sailed along the East Coast and up into the Thames towards the 'Chiefest Port' had few, if any, navigational aids to guide them.

Defoe when making his tour of Great Britain wrote of the great quantities of fish caught off the Essex coast and remarked: 'On the Shoar beginning a little below Candy' [Canvey] 'Island or rather below Leigh Road, there lies a great Shoal or Sand called the Black Tayl, which runs out near three Leagues into the Sea due East . . . From this Sand, and on the Edge of Shooberry, before it, or South-West of it, all along, to the Mouth of Colchester Water, the Shoar is full of Shoals and Sands, with some deep Channels between.'

A glance at the chart shows just what he meant. The funnel-shaped estuary of the Thames—its boundary is said to be a line from the Naze north of Walton to the North Foreland—contains a whole series of long narrow sandbanks running in north-easterly and south-westerly directions, with channels between them and other channels across them.

Closest channel to the shore is the Wallet, running along by the modern holiday resorts of Walton and Clacton, sheltered to seaward by a strip of hard sand bearing the name of the Gunfleet, a bank which was for many years marked with a lighthouse erected on screw piles, now standing lightless above the lapping waves.

To seaward again is the Swin or King's Channel, a channel used in the past by much of the collier and barge traffic bound from the north to the Metropolis. 'Owing to the nature of the navigation in East Swin and to a want of due caution, more wrecks have occurred in it than

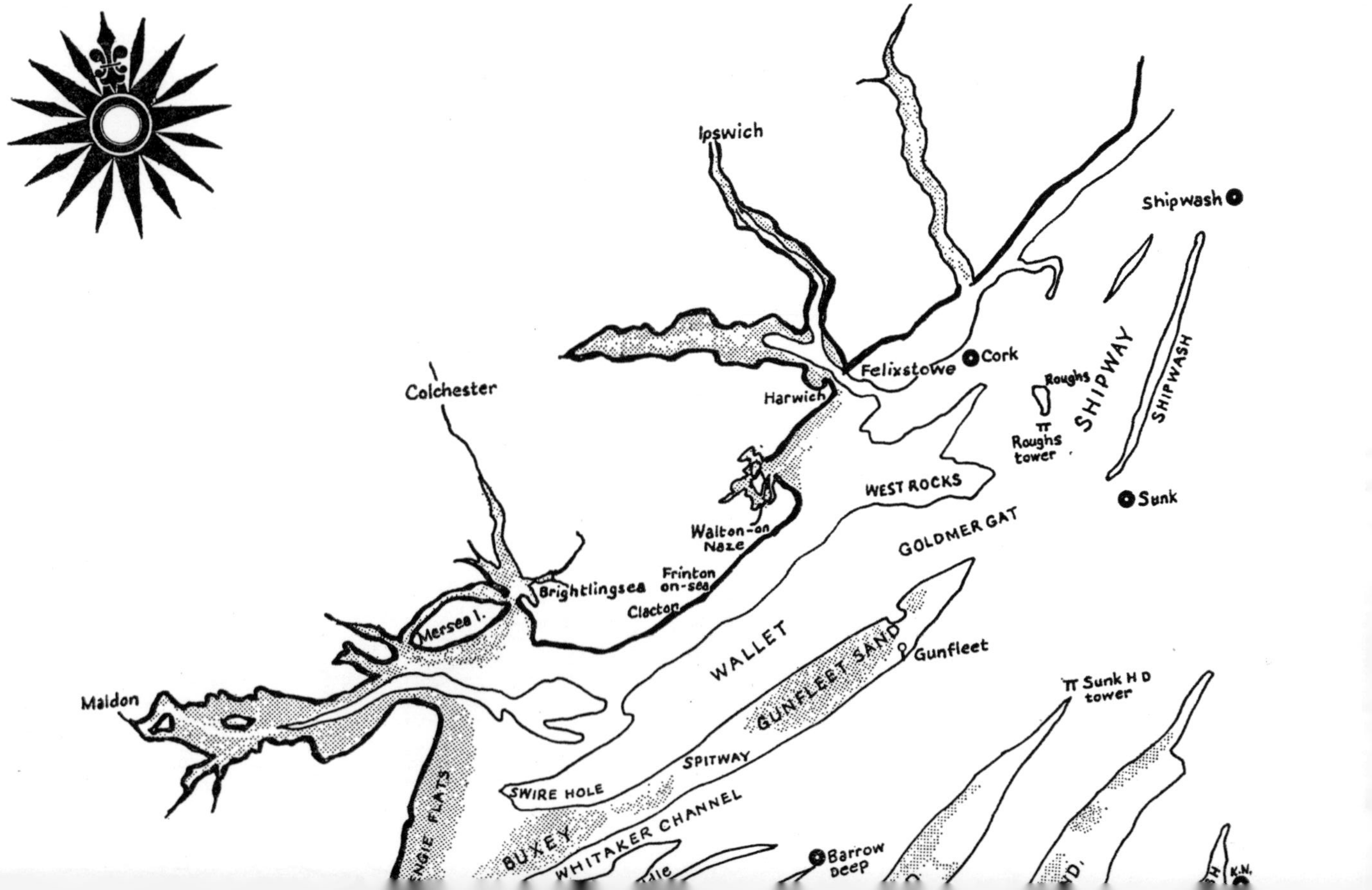
Ipswich
Shipwash
Cork
Felixstowe
Harwich
Roughs
Roughs tower
SHIPWAY
SHIPWASH
Sunk
Colchester
WEST ROCKS
Walton-on Naze
GOLDMER GAT
Frinton on-sea
Brightlingsea
Clacton
Mersea I.
WALLET
GUNFLEET SAND
Gunfleet
Sunk H D tower
Maldon
SPITWAY
SWIRE HOLE
WHITAKER CHANNEL
BUXEY
Barrow Deep
K.N.

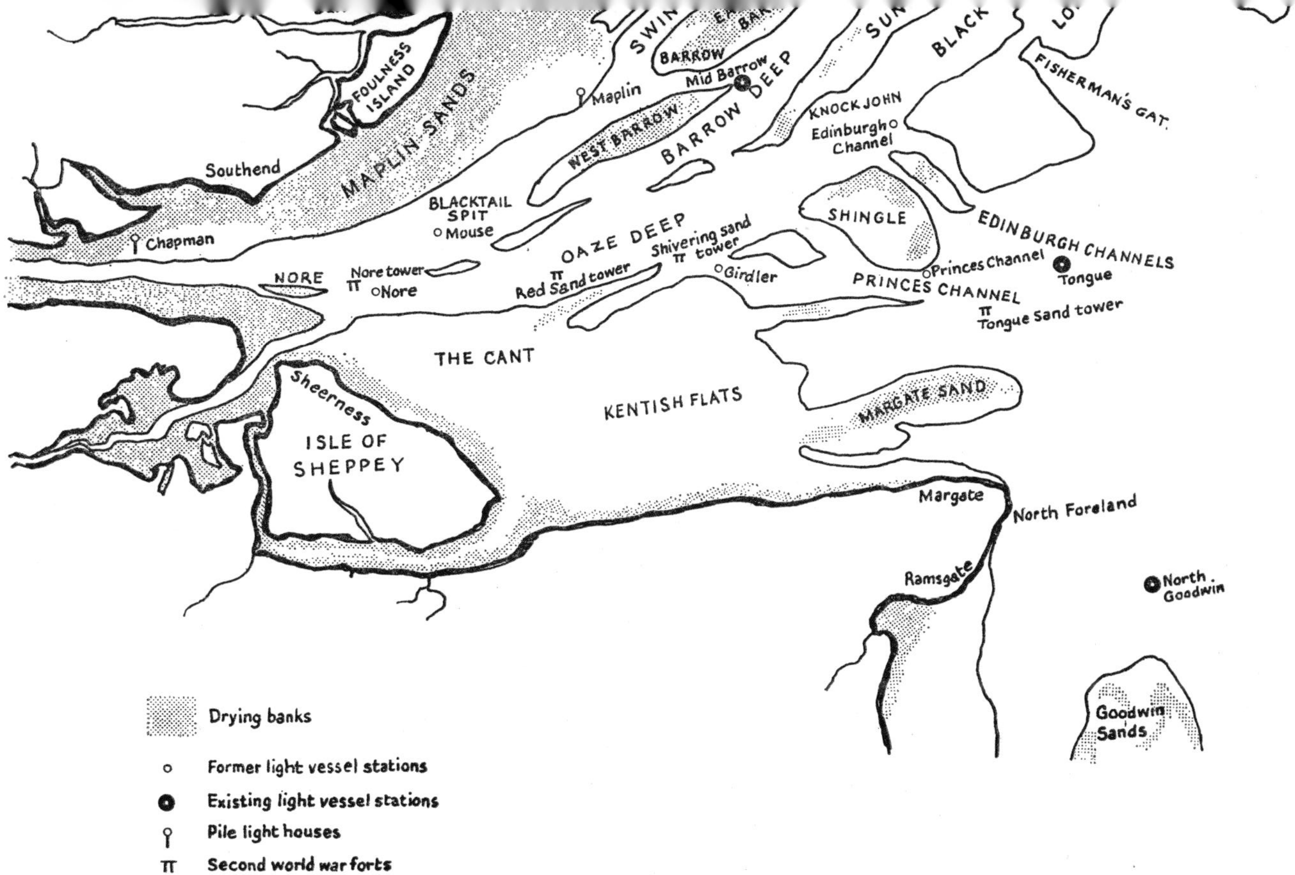
FOULNESS ISLAND
MAPLIN SANDS
Southend
Chapman
Maplin
BARROW
Mid Barrow
WEST BARROW
BARROW DEEP
KNOCK JOHN
Edinburgh Channel
FISHERMAN'S GAT.
BLACKTAIL SPIT
Mouse
SHINGLE
EDINBURGH CHANNELS
OAZE DEEP
Shivering sand tower
NORE
Nore tower
Nore
Red Sand tower
Girdler
Princes Channel
Tongue
PRINCES CHANNEL
Tongue Sand tower
THE CANT
KENTISH FLATS
MARGATE SAND
Sheerness
ISLE OF SHEPPEY
Margate
North Foreland
Ramsgate
North Goodwin
Goodwin Sands
Drying banks
Former light vessel stations
Existing light vessel stations
Pile light houses
Second world war forts

upon any other portion of the eastern coast,' stated the Admiralty sailing directions for the North Sea early this century.

Then come the Barrow and the Sunk sands, between which is the Barrow deep. Further to seaward, beyond the Black deep, lies the Longsand. Furthest out, about twenty miles south-east of Walton, is the Kentish Knock, scene of some tragic wrecks and some spectacular rescues. Near the coast to the south of the Gunfleet, of which it forms an extension, lies the Buxey, which has also taken its toll of shipping.

To the southward off the mouth of the Thames lie the Shingles, Girdler, Tongue and Margate sands, while closer in is the Middle sand and right in the mouth of the river itself, between Southend and the Isle of Grain, is the Nore sand. There are also sandbanks along the shore, as Defoe mentions, extending out for as much as four miles from the sea walls which protect the low-lying coastal marshland. These include the Dengie flats and the Ray sand, the Foulness and Maplin sands, and, in the mouth of the Thames itself, the Southend flats and Chapman sands on the Essex shore and the Yantlet flats and Blyth sand off the Kent coast.

All these sands and the channels between them are constantly changing. As an instance of this it might be mentioned that the Edinburgh channels, a bifurcated passage between the Longsand and the Shingles, were formerly one, known as the Bullock channel. But the formation of Shingle patch in its centre around 1882 split it in two, the North and South Edinburgh channels.

Such was the importance of the trade routes along this coast that during the seventeenth century Trinity House was placing buoys in the main channels to supplement the few beacons which already guided shipping, and by the time Pepys became secretary of Trinity House there were buoys all along the Essex coast, the north buoy-keeper being based at Harwich.

It was in these waters that the first lightship made its appearance, at the Nore, roughly midway between the Essex coast and the Kent shore, in 1731. A 'floating light' had been suggested fifty years before, but had been damned by the Elder Brethren, who loftily refused to allow such vexatious proposals to be taken seriously. Within the next century and a half, however, the number of lightships off the Essex coast and in the estuary had grown to a dozen, and these were complemented by four pile lighthouses, on the Gunfleet and Maplin sands off the coast and on the Chapman sands and Mucking flats in the estuary.

Yet even in 1910 the *North Sea Pilot* found it necessary to warn that 'in thick or foggy weather, or in strong winds or gales with snow, too much precaution cannot be exercised; and, when at all doubtful as to the vessel's position, it is far safer to anchor and wait until the fog or mist clears sufficiently to allow objects at a distance of two or three miles to be seen, rather than by proceeding to risk running on any of the banks or a collision with another vessel'.

Disregard of this warning led to the loss of many ships, large and small, and the *Pilot* also records that 'The frequent casualties on the sands in the Thames estuary have caused the Trinity House to adopt special signals on board the lightvessels, and at the lighthouses, to facilitate the diffusion of the intelligence of a vessel being in distress, and of the locality of the mishap'—government jargon is nothing new.

The story of lifesaving off the Essex coast is not simply the story of lifeboats, because, as in Norfolk and Suffolk, there were professional salvagers who, however badly their names were blackened at times, carried out some epic rescues while going about their business. So successful were they that it was not until 1876 that the Royal National Life-boat Institution stationed a lifeboat anywhere on the Essex coast. Since that time, however, the various lifeboats have carried out a great deal of work and have effected some very creditable rescues.

In writing the story of these stations it has seemed best to concentrate to some extent on the most interesting aspects of each station while at the same time attempting to produce a balanced history. It seemed that the experimental work done at Harwich, first with steam lifeboats and later with motor lifeboats, was the most interesting aspect of the story of that station; that the rivalry between the Institution lifeboat and the volunteer lifeboat between 1894 and about 1911 provided some of the best stories from Walton; that Clacton's preoccupation with crashed aircraft in two wars provided a highlight there; and that the most interesting part of Southend's history was the work of the lifeboat in the war of 1939-45. It is to be hoped that no reader will be disappointed to find a certain concentration on what the author has felt to be the most interesting periods in the history of the various stations.

Although there are about 40 miles of coastline between Harwich and Sea Reach—not allowing for the inlets of the Colne and Blackwater and the Crouch—there are few places really suitable as lifeboat stations. Between Clacton and Southend there are no coastal towns or villages at which a lifeboat crew could be mustered, except for Burnham-on-Crouch, which is five or six miles upriver and so relatively useless as a lifeboat station, though it does now have one of the many inflatable inshore rescue boats used mainly for dealing with summer yachting accidents and the like. Hence the distribution of stations, with Harwich, Walton and Clacton set close together on the northern section of the coast and Southend almost thirty miles away in the mouth of the Thames. Although set so close together, the lifeboats based at Walton and Clacton have a considerable area to cover, and even today with the extended range of modern motor lifeboats they both have a role to fulfil—though with the re-establishment of the Harwich station one wonders if some pruning is, perhaps, imminent.

In writing the earlier part of the story much information has been gleaned from the pages of local newspapers, primarily the *Essex Standard*—now the *Essex County Standard*. Official records have also been consulted where available, but it should be pointed out that in the case of the so-called private lifeboats no records have survived and the story has had to be told almost entirely by reference to contemporary newspaper accounts, supplemented by such as can still be remembered by members of the older generation.

Most of the wrecks in this area occurred on the outlying sands, often as much as twenty miles from shore, and it has been difficult to find photographs of lifeboat services and wrecks. The men of the Walton lifeboat now carry a cine camera when they go on service, and films of their work have appeared on television, but years ago there was no opportunity for photographic work in a pulling and sailing lifeboat and no photographer thought of going off in a boat to take pictures of a wreck. After all, these were not unusual occurrences, happening every week—if not almost every day—and anything so commonplace did not attract either artist or photographer. Only occasionally when a wreck such as that of the *Deutschland* attracted unusual attention were engravings included in the illustrated periodicals of the day. It would have been amusing to have found a painting of the rescue of the crew of the little Ipswich sloop *Sarah* by horse and cart at Little Holland in 1836, but this stirring act seems to have gone unpainted though not, happily, unreported.

It has, however, been possible to find photographic records of many of the lifeboats and their crews, and my thanks are due to all those who loaned old pictures for copying or even provided me with copies. My thanks are also due to the many who helped by supplying information and to those who combed my writings in an effort to ensure accuracy. Mistakes are almost certain to have crept into a book such as this, and I must apologize in advance for them.

In particular I have received much help and very practical encouragement from Grahame Farr, of Bristol. Patrick Howarth, Christopher Elliott and other members of the R.N.L.I. staff have gone to considerable trouble to supply information, as have the librarians and staff at Colchester, Southend and Tower Hamlets Public Libraries and Essex Record Office and the pier and foreshore manager at Southend, G. S. Dawson. My thanks are also due to the editors of the *Essex County Standard* and *East Essex Gazette* for permission to use photographs taken for those newspapers.

The sketches of Essex smacks on pages 15, 21 and 22 are by John Leather, and the map on pages 8 and 9 was specially drawn by Keith Mirams.

I should like to thank, too, V. G. Scamell, John Leather, Frank Putman Ltd., D. Royston Booth, Norman Harrison, J. H. Turner, Jack Mitchley (of the Port of Lowestoft Research Society), Lieut.-Cdr. D. J. Wilford, B. Peachman, B. Norman, A. Horton, Mrs. S. H. Cooper, R. Saunders, Mrs. Lloyd, F. Z. Claro, C. Schofield, G. H. Cooke and many others who have helped in one way and another. Without their help this book could never have been written.

R. W. MALSTER
Colchester

December 1967

ABBREVIATIONS

The following abbreviations have been used in the Boat Record tables to denote types of lifeboat:

IRB =	Inshore rescue boat	SR =	Self-righting
L =	Liverpool	W =	Watson
(M) =	Motor	N&S =	Norfolk and Suffolk

1

THE SALVAGING SMACKS

SOME OF THE most important trade routes in the world pass along the Essex coast, and much of the shipping approaching the port of London uses the narrow channels which lie close to the county's shore. There is deep water in the Swin, in the Barrow deep and the Black deep, but the channels are bordered by sandbanks, some of which dry at low water, while others lie for ever just beneath the lowest of spring tides. All are equally ready to trap and hold any vessel, large or small, which gets ever so slightly out of its reckoning or which cannot make headway against an easterly gale.

Once, Indiamen brought home spices and silks to line the pockets of London's merchants—if they did not come to grief among the sands; collier brigs and schooners brought coal from the Tyne for London's grates; small barques from the Baltic brought timber and deals for London's shipyards and timberyards; full-rigged ships and barquentines of no great size distributed goods all over the continent of Europe from the world's greatest entrepôt; big ketch barges sailed coastwise and to the nearer European ports with any cargo they could pick up; little stackie barges brought hay for the capital's cab-horses and dray-horses and returned to Essex farm quays with odoriferous cargoes of manure—the street-sweepings of the city; and the packets took passengers and mails from Harwich to the mainland of Europe. These were indeed busy waters, frequented by traders of all kinds as well as by the fishermen of Barking, Leigh, Brightlingsea, Wivenhoe and Harwich.

It is not surprising that there should spring up a race of professional salvagers in such an area. For men who were skilful seamen and knew no fear of the sea the pickings could be good, though the risk was great. In their fine weatherly smacks the Swin Rangers could keep the sea in almost any weather. They knew the sands and the channels between as well as a landsman knows the streets and lanes of his town or village—usually better—and they made use of their skill and knowledge to pluck many ships and many lives from the sands. When there was no chance of saving the ship they often turned to plunder, but their daring and skill in saving life made them a respected if sometimes not particularly law-abiding community.

The Essex smacks were fine, weatherly cutters quite different from the little dredging smacks which can be seen today converted into yachts, for the first-class smacks were anything from 55 to 70 feet long. Many of them were built around the middle of the last century when the coming of the railway stimulated the local fishing industry and gave

The Essex smacks were built for speed as well as seaworthiness—the New Unity, *built at Rowhedge by Peter Harris, alongside the quay at low water.*

promise of further development which, however, never came. Aldous of Brightlingsea built no fewer than 36 of these smacks between 1857 and 1867, while Harris at Rowhedge and Harvey, Barr and Husk at nearby Wivenhoe built many more, others being constructed on the Blackwater.

The Rowhedge *Aquiline*, mentioned later in this chapter, can be taken as a typical example of a first-class smack from the Colne. Built by Harris in 1867, she measured 21 tons and was 65 feet long, with a 15 foot beam and a draught of 8 feet 6 inches when loaded. She carried a sail area of about 2,000 sq.ft. of working canvas, the mainsail having a 45 foot boom.

Such, then, were the powerful cutters in which the Essex men put to sea in search of oysters and scallops—and salvage. Owned in the villages of Brightlingsea, Wivenhoe, Rowhedge and Tollesbury, they were registered at Colchester. They sailed down Channel to exploit the rich oyster beds off Jersey and round to the westward to work the beds in Swansea and Cardigan Bays; they exploited, and ultimately ruined by overfishing, the oyster beds found off the Norfolk coast; they crossed the North Sea to dredge on the Terschelling banks, off the Dutch shore; and they would go 'stowboating' for sprats on a large scale off the Essex coast.

In 1823 the master and two men of the Brightlingsea smack *Prince Coburg* were lost when attempting to board a dismasted vessel in the Swin, and the *Colchester Gazette* referred to 'so valuable a body of men as that to which the unfortunate sufferers belonged'. Many gave their lives when heavy seas overturned their little boats as they pulled from their smacks to vessels in distress or when the boats were swamped by the breakers on the sands.

Stories of plundered vessels are legion. The London brig *Lochiel* ran on the Maplin just before Christmas 1840, in a gale from east by south and thick snowstorms and was stripped of stores and gear before the eyes of her master and crew. Capt. Benjamin Garrick, the master, gave an account of the incident later, telling how at daybreak, after the vessel had been a night on the sands, eight smacks came up and the smacksmen boarded the brig. Finding that their efforts at jettisoning the cargo of coal and pumping her out were unavailing, Capt. Garrick decided to send his crew off in the smacks' boats.

'We were told by the smack men that, in consequence of their boats being so small and the weather so bad, they could take only a few of their most valuable things; which was done, and put into the boats, leaving our other things on board, in the hope of the weather moderating enough to afford us a chance of recovering them. But as soon as we had placed our things in the boat, the other smacksmen that were on board went down below, breaking open the chests and lockers, and taking things out of them, and loading their boats, with the exception of Capt. E. Eagle, master of the smack *George and Eliza*, of Colchester, and Capt. Andrews, of the smack *Fair Traveller*, of Colchester, who would take nothing but what belonged to the men they had to take off, and endeavoured to take as many of my things with them as possible. They frequently reprimanded the other men for taking and appropriating to their own use the clothes of the shipwrecked sailors.'

Capt. Garrick went aboard one of the smacks, which proceeded into

Sheerness for shelter. When they returned to the wreck next morning he found her stripped of everything except the lower and topmast standing rigging.

Then there was the case of the Newcastle snow *Antoeus* which hit the Gunfleet one November day in 1844 when nearing the end of a voyage from Finland with deals. Her anchors were let go, but the cables broke and the salvors' anchors also failed to hold her. The master and crew decided to abandon the vessel after some of her materials had been put on board the smacks which stood by.

'Upon the master's return in the afternoon of Saturday it seemed the wreck had suffered great pillage, but that it had been put a stop to by the appearance of the Deputy Vice-Admiral (J. G. Chamberlain Esq. of Wivenhoe) who was then on board the snow . . . Exertions will no doubt be used by the authorities and others interested in the ship to bring to justice the parties guilty of plundering the ship."

But that was not the worst of it. A few days after the *Antoeus* had been stranded some of the salvagers lit a fire in the cabin to warm themselves. The fire was perhaps too big for the stove, or maybe it was not in the stove at all, and the snow was burned to the water's edge.

Yet there was another side to the picture. When the fishing smack *New Maria* of Ramsgate was lost on the Longsand in January 1842, the three survivors of the crew of five were rescued by John Lewis in the smack *Endeavour* of Harwich after they had spent 20 hours in the rigging. In November the same year the same John Lewis, a member of a Harwich seafaring family, went to the assistance of the crew of the *Hamilton*, a 'very fine vessel' of upwards of 300 tons, which went on the Gunfleet in thick weather and a heavy gale. But before he could slack his boat alongside the wreck the *Hamilton's* mainmast, in the rigging of which the crew of ten were clinging, went overboard. Only one man was picked up alive.

Another salvaging smack owned by the Lewis family was the *Aurora's Increase*, a 22-tonner. Two of the crew of this vessel were summonsed at Harwich Town Hall in 1851 by the Receiver of Droits of Admiralty 'for unlawfully retaining divers stores and materials, viz. the mainsail, a topsail, squaresail and a quantity of rope which had been picked up by them at sea'. Although they had landed the gear almost as soon as they came into port and declared their intention of reporting it as soon as the Receiver's office opened next morning, they were convicted and fined fifteen guineas each, double the value of the articles seized.

The magistrates' decision created a sensation in the court, which was crowded with members of the seafaring fraternity. John Lewis, who told the court he was both master and owner of the *Aurora's Increase*, said that he had spotted the Sunderland brig *Loyalty* on the Whiting sand and had manned his boat and boarded her in a gale. 'After being on board about six hours I got her off at dark, with 4½ feet water in her; brought her up, she making water fast; lost one of our boats; took the only remaining one we had to row to the smack to get the crew off, as the ship was sinking; my brother, the captain, and crew of the brig remaining for me to take them off. While I was bringing my smack up some had jumped into the ship's longboat and rowed away leaving my brother, the captain and three or four others without a boat or the least means of escape, the brig gradually sinking. I succeeded in the dead of night in reaching the brig; the water was then over her decks, and I took my brother, the captain and others off the rigging, and in five minutes she sank. The next day I recovered the sails . . .'

Whatever the view taken by the Law, the Shipwrecked Fishermen's and Mariners' Royal Benevolent Society did not regard the smacksmen as criminals. They sent £20 for the masters and crews of the smacks *Tryall*, *Aurora's Increase* and *Wonder*, William and John Lewis and George Barrett, in recognition of their 'brave and humane conduct'.

Later the same year the society awarded the crews of the *Tryall* and the *Aurora's Increase* £5 and voted silver medals to William Lewis, Thomas Tye, William Soder, George Wyatt and Joseph Day after they had rescued the crew of the French brig *Etoile de la Mer* from the Longsand.

To show how busy the east coast was at that period, it may be noted that in one day of February 1843, nearly 200 ships came into Harwich harbour to shelter from the easterly gale that was blowing. Most of them had lost an anchor or had sustained other damage.

Every blow brought its list of casualties in those days when the Swin was crowded with sail. One week's report in the *Suffolk Chronicle* can be taken as an example: one vessel had foundered at anchor off the West Rocks, near the entrance to Harwich harbour; a brig was reported on the Sunk; another vessel with the suggestive name of *The Ark* had been assisted into Harwich leaking after having been on the Gunfleet; and a further vessel was assisted in after having been on the Shipwash, off the Suffolk coast; while yet a further vessel was brought in also having been aground. Two days later the *Britannia* was brought in, having been abandoned.

In fog and rough weather the sands could indeed be a death trap for shipping, and many a rescue was carried out by the smacks when other vessels were running for shelter or, if already in port, looking apprehensively to the security of their moorings. But even the Swin Rangers found it beyond their powers to help on some occasions, one being when the brig *Traveller* got on the Gunfleet in a howling gale in February 1843. She went ashore on a Saturday night, and was already fast breaking up the next morning when she was spotted by two Colchester smacks, the *New Gipsy* and the *Atalanta*. The crew took to the rigging as the brig settled in the sands, and their suffering must have been intense. Throughout Sunday the smacks stood by, but the smacksmen saw that no boat could live among the breakers and there was no way of approaching the wreck until the weather improved.

The smacksmen could only encourage the shipwrecked men by hoisting a flag as a signal that they would not abandon them. All through that day and the next night the smacks stood by, and then on the Monday morning, the weather having moderated just a little, two boats were manned and pulled, not without much effort and difficulty, to the wreck. Six men were taken from the rigging by the *New Gipsy*'s boat and four by the *Atalanta*'s.

The breakers were still crashing on to the sands and it seemed hardly possible for anything so small and frail as a smack's boat to survive in the confused and broken water; the crews of the two boats could not agree which was the safest way to leave the sands, and in the end one dropped down to leeward and the other was pulled into the Swin, where she fell in with the s.s. *Gazelle*, and put her six men on board. The steamer lay to and the four men saved by the *Atalanta*'s boat were also transferred to her from that smack. The men had had a terrible experience; they had suffered so much from exposure during more than 30 hours in the rigging that their limbs were so swollen that their clothes had to be cut away.

One of the smacksmen who took part in the rescue of the *Traveller*'s crew in 1843 was John Glover, a Rowhedge man who later had his own smack, the *Increase*. Another of his rescues was from the *Dane of Argyle*, of Perth, which was wrecked on the Sunk sand. The crew were all washed away with the exception of three men who clung to the rigging in desperation for many long hours. When the smacksmen fought their way to the wreck and got two of them into the boat they told the smacksmen their mate was so nearly dead it was no use taking him, but this did not suit the stout-hearted Glover. 'Never mind, dead or alive, we'll save him,' he replied—and save all three he did.

He was said to have been responsible for saving nearly 300 lives; his most notable rescue was probably that carried out in 1870, when he took thirteen men from the Prussian barque *Carl Agrell*, which had run on the Longsand on December 23rd when running before a heavy gale in blinding showers of snow. Rolling seas swept over the stranded vessel and forced her further on the sand, and two men who attempted to launch the boats were lost when these were swept away.

Glover came on the scene in the *Increase*, apparently with a double crew, for the crew of the smack *Reindeer* were aboard as well as his own men. But even a double crew was no match for conditions prevailing at the time, and twice during the day attempts to reach the wreck in what the *Essex Standard* described as a lifeboat—perhaps the smack's boat had been modified with barrels beneath the thwarts or a cork wale—were beaten back by the fury of the storm.

Night came on and the German crew had to lash themselves in the rigging, expecting every moment that the seas which broke across the decks would take the masts overboard. The *Increase* stood by all night, and at daylight the boat was again manned. By almost superhuman efforts the smacksmen got alongside the barque and took off the thirteen half-drowned and exhausted men, who were supplied with what dry clothing the Rowhedge men could muster and were landed at Brightlingsea.

More than a year later Glover received a presentation telescope from Kaiser Wilhelm, 'king of Prussia and emperor of Germany', in recognition of his bravery on this occasion. It is still in the possession of his great-great-grandson. After risking his life for many a year among the sands of the Essex coast, Glover died at his home in Harwich at the age of 82.

Another smacksman who lived to a ripe old age in spite of the risks he ran was Tom Barnard, probably the most famous Swin Ranger ever to sail from Rowhedge. Many are the tales told about him—not all of them true.

Barnard was already no youngster when he suffered shipwreck himself, losing the smack *New Unity* in a December gale—not on one of the sands he knew so well but on a fort at the entrance to the Medway.

There were few occasions indeed when the smacksmen could be accused of not giving assistance, but the name of the salvagers was blackened—most unfairly, as it later appeared—in both British and

German newspapers over the *Deutschland* affair in December 1875.

The *Deutschland* was a Hamburg-Amerika liner of 2,898 gross tons built on the Clyde by Caird in 1866. When she left Bremen for New York on one voyage towards the end of 1875 she was carrying 123 passengers, mostly emigrants leaving their homeland for a new life in

The New Unity *on the Skilling grounds*

the New World, and a crew of 105. She was proceeding dead slow through rough seas in the early morning darkness, made more dense by thick, swirling snow, when she grounded on the Kentish Knock about 5 a.m. on December 6th. The engines were put full astern at once to draw her off, but within seconds the screw broke off, and then the steamer was helpless for no further attempt could be made to get her off under her own steam.

Two boats were lowered, and one of them came ashore at Sheerness with three men, two of them dead from exposure. The other boats could not be launched because of the weather, and ultimately they were all washed away by the heavy seas.

During the day efforts were made, by throwing cargo overboard from the fore hold, to keep the *Deutschland* from getting broadside to

The Aquiline *in very different weather from that prevailing when the* Deutschland *was wrecked.*

the seas, and the pumps were kept going to keep her clear of water. But as darkness came on and the tide rose the decks were awash. Crew and passengers had to take to the rigging, although Captain Brickenstein remained on the bridge until the seas washing across the vessel forced him to take to the rigging with the others. And there the crew and passengers had to remain for many hours.

Among the smacks which put to sea to attempt to give assistance was the first-class cutter *Aquiline*, owned by Harry Mills Cook. They could get nowhere near the liner, which could not be seen through the snow squalls as she lay in the shallow broken water on the sands. Then the *Aquiline* sprang her boom and, crippled, had to put back to Walton Backwaters, the only shelter that could be reached to leeward. Her crew fished the boom, but it blew so hard that there was no hope of the *Aquiline* beating out against the E.N.E. gale.

The crew put Harry Cook ashore at Oakley and, dressed in seaboats and pilot gear, he walked across the marshes to the road and into Harwich as quickly as he could to give news of the wreck. Footsore and tired he must have been when he got to Harwich Coastguard station and told the Coastguard of the wreck, only to find that they had already had news of the boat which had been blown ashore at Sheerness. He was able, however, to tell them where the wreck was, which they had not learned from the *Deutschland* survivor.

Leaving the Coastguard station, Cook went down to Walter Watts' tug *Liverpool* and went out on her when she left harbour at first light. When she arrived in the vicinity of the wreck about noon the tugmen found that the *Deutschland* was beginning to break up under the impact of the heavy seas, which were sweeping overboard many of the passengers and crew. The tug could not at first get alongside and Captain John Carrington brought his vessel up in three fathoms some distance off. The tug's boat was launched and three boat-loads of women and children were ferried from the wreck to the tug. As the tide flowed, the *Liverpool* was able to get alongside and took off the remaining people. Altogether 155 people, including 21 women and children, were saved by the tug, and the little *Liverpool* must have resembled a pleasure steamer as she paddled back to Harwich. But it was nothing of a pleasure trip. 'The rescue was effected under the greatest difficulties I've ever known, although I've been at sea altogether nearly 50 years and out of that time engaged for 40 years in salvage work,' said the tug skipper later.

Next day when the *Liverpool* went out again there were fourteen smacks and luggers gathered round the wreck, according to contemporary newspaper reports.

The *Liverpool*'s efforts resulted in the rescue of a large proportion of those aboard the ship, but 44 passengers and 20 of the crew were lost before the tug could reach the scene. Great was the outcry, and accusations were levelled at the Essex men that they had sat in port without doing anything to give help long after they knew the vessel was ashore.

But after the inquiry *The Times* had to admit that 'the Harwich men displayed none of the remissness the first accounts of the disaster seemed to imply'.

Of course great play was made of the fact that there was no lifeboat at Harwich. For many years it had been thought better for the Suffolk and Kent lifeboats to deal with wrecks on the Essex coast; although on one occasion in 1876 when a foreign barque went on the Shipwash and the news was known in Harwich within a few hours it did not reach Ramsgate, where a lifeboat was based, until more than 24 hours later. In fact it had always been agreed that the smacks did more good work than could be accomplished by any lifeboats, and every inspecting officer of Coastguard who had been called on to report officially on the subject in the twenty years preceding the *Deutschland* disaster had stated that no lifeboat was needed at Harwich.

By the end of the century the day of the big Essex smacks was drawing to its close. One might think that smacks which could ride out such conditions as those the salvagers operated in could survive any weather, but the sea took its toll, and a terrible toll it was. The blackest year of all for the little village of Brightlingsea was 1883, when two disasters followed one another within nine months. Three of the Brightlingsea fleet of fifteen vessels that had been working on the 'Skilling' grounds off the Dutch coasts failed to return after a very severe storm in March, and in a December gale two smacks were overwhelmed on the Terschelling banks.

What it was like to sail through such a storm as that in which the *Recruit*, *Conquest* and *Mascotte* were lost without trace was described later by the master of the *Express*, a 26-ton cutter built at Brightlingsea in 1855. She was struck by a heavy sea which carried away the mainsheet, broke the boom and washed overboard the helmsman. The rest of the crew tumbled out in time to see their friend being swept away; but with the smack crippled by the damage she had received, they could do nothing to save him. They spent three exhausting hours clearing the broken boom, which was threatening to hole the smack, and in getting the mainsail below and setting a jib abaft the mast.

A few hours later another violent sea struck the vessel, carrying everything before it except the mast—luckily the helmsman had lashed himself at the tiller and was safe. The smack was forced on to her beam ends, the stove was thrown into one of the bunks, and 11,000 oysters, together with the dredges, warps and other gear, were shot into the leeside of the hold right under the deck. She therefore could not right herself.

'To us who were below deck it fared to be some time before the deck could be reached—the side of the cabin answered for the floor, so that we could not reach the hatch,' said skipper George Butcher.

But they managed to clear up and trim the smack while lying to two dredges hurriedly cast overboard to bring her head to sea. When the weather improved a little they managed to sail her home under jury rig.

The 1883 disasters led to something of a decline in the Terschelling fishery, and poisoning scares in the early years of this century brought a decline in the demand for deep-sea oysters. At the same time the big smacks of Harwich were disappearing as the focus of the cod fishery moved to Hull and Grimsby. Some of the Colchester smacks were sold away from their home ports, and only a few kept going through the first World War.

If the nineteenth century had been the age of salvaging smacks, in the twentieth the sea was being left clear for the lifeboats.

2

HARWICH

'THE PORT of Harwich, owing to its general depth of water, its extent, the shelter it affords, and immediate communication with the sea, is one of the most valuable on our eastern shores; and although the rivers Thames and Humber afford shelter by running far up them, yet Harwich, from its easy access by night or day, in all weathers, and in all states of the tide, is the only harbour of refuge, properly so called, on the East coast of England.'

That was the view of Capt. John Washington, who had been engaged in surveying work in the North Sea, when he made a report to the Secretary of the Admiralty in 1843. He recalled, too, that Harwich had been of considerable importance as a naval base, that warships had been built there, and that it was also the home of a considerable fishing fleet.

At the time he was writing Harwich was in decline, but it had earlier been of no small consequence as a packet port. During the seventeenth and eighteenth centuries the packet service to Helvoetsluys had been a source of profit to the traders of Harwich, but the development of steamships and the building of Britain's network of railways resulted in a temporary loss of trade. The coming of the railway to Harwich in 1854 restored the place to its former position as a port for the Continent, yet even today the town has not regained its former eminence as a seaport.

Although it was not until 1876 that the R.N.L.I. sent a lifeboat to Harwich, there had been earlier short-lived lifeboat ventures there. The story is far from clear, but it would seem that the smacks did far more work than these early lifeboats.

The wreck of a schooner on the Cork sand with the loss of nine men in 1820 led to moves to establish a lifeboat at Harwich. There were already two lifeboats on the Suffolk coast, but the *Colchester Gazette* lamented that the establishment of lifeboats on the Essex coast had been neglected.

People at Harwich wrote to Captain George Manby, the inventor of the life-saving mortar and an indefatigable worker in everything that savoured of life-saving, and his advice was considered at a meeting a few days before Christmas. At this meeting, held at the Moot Hall in Colchester, it was decided to have two lifeboats, one at Harwich and the other at Brightlingsea.

The Harwich gentlemen agreed to join with those who had met at Colchester, and on the first day of 1821 the Essex Lifeboat Association was formed and a lifeboat was ordered from George Graham, who had the Naval Yard at Harwich. He showed a model of the boat at another meeting the following month, and it was decided to begin the building of the Harwich boat immediately, but the building of the second was 'unavoidably deferred until the benevolent object of the Association receives a more general support from the friends of humanity'.

Meanwhile another subscription was being raised at Ipswich for a lifeboat to be stationed on the Suffolk side of Harwich harbour at Landguard Fort. This boat was designed by Captain R. H. Gower, a former East India captain who was responsible for a number of unusual vessels, and was built by Jabez Bayley, the Ipswich shipbuilder.

The Harwich lifeboat was a big heavy craft, but that at Landguard was a light one after the form of a whaleboat, capable of being transported from one part of the beach to another. Double-ended and rigged with two very short masts on which were set spritsails, she had two self-delivery tubes for clearing herself of water and gained her buoyancy from 11 cubic feet of cork and 50 cubic feet of copper air cases.

In the autumn of 1821 the Orwell lifeboat, as she was known, was rowed across the harbour to be present at the launching of the Harwich lifeboat by Mrs. Anthony Cox, the mayoress of Harwich, who named her the *Braybrooke* in honour of the Lord Lieutenant of Essex. She should have been ready the previous month, but the ceremony had had to be postponed until September 11th.

Unfortunately there was not enough money in the Essex Lifeboat Association's funds to pay for a boathouse and the *Braybrooke* had to lie afloat in the harbour. It was found that in heavy weather nobody could reach the boat, so it was not surprising that early in 1822 the Association was asking for annual subscriptions to be paid promptly as 'funds are wanted to man the boat and enable her to act, should any case of peril call for her assistance'.

Such a 'case of peril' occurred one day in May 1822, eight months after her launching, and the *Braybrooke* was used for the first time. A vessel went on the West Rocks in an easterly gale and the *Braybrooke* and other boats from Harwich went to her assistance and got her off successfully. This was not her only salvage operation, for some time later the Commissioners of the Navy awarded her crew £10 for assisting H.M. cutter *Surly* when she ran ashore.

The Orwell lifeboat was launched one February day in 1824 after her crew had been summoned by the firing of guns and burning of 'blue lights' by the garrison at Landguard Fort. The brig *Malvina* was found aground on the Platters off Felixstowe and was assisted into Harwich harbour. The *Braybrooke* was in action the next year, when she went out to a brig on the West Rocks, only to find the crew had abandoned her and had been picked up by another vessel.

Sometimes, however, the inactivity of the lifeboats led to an outcry. When the brig *Elegant* was wrecked on the Rough sand in 1822 it went unseen for so long that only one man, the captain, was rescued. There was great trouble at Landguard, where the lieutenant governor of the fort ordered an investigation into the reason why none of his garrison had told him of the wreck.

The Orwell lifeboat was not a great success and was eventually sold for conversion into a yacht. What became of the *Braybrooke*? Nobody seems to know, for local newspapers are quite silent on the subject of lifeboats after 1825.

The Admiralty stationed a lifeboat on Landguard Point about 1845, but there is no trace of her ever doing any useful work and for the most part it was left to the smacksmen to save the crews of vessels wrecked on the sands off the Essex coast.

After the disastrous wreck of the *Deutschland* on the Kentish Knock, a lifeboat was sent to Harwich by the Royal National Life-boat Institution. This boat, the *Springwell*, was towed from London to Harwich by a steamer in January 1876, and by the time she was officially handed over on the anniversary of the *Deutschland* disaster had already been out on service. By the end of 1879, she had been out fourteen times and had saved 39 lives.

One of her major rescues was performed in February 1879, when she saved 23 men from the Liverpool barque *Pasithea*, aground on the Longsand. When the *Springwell* and the tug *Harwich* reached the barque the tide was too low for them to approach, but as the water rose the lifeboatmen boarded her and the lifeboat took towropes from the barque to the tug.

Time after time the towropes broke as the paddle tug strained to haul the vessel off the hard-gripping sand. Ballast was thrown overboard to lighten her, but still the vessel remained fast, bumping heavily in the high seas then running. About 9 p.m. the order was given to abandon her. As the barque's crew and a salvage crew of ten smacksmen were being transferred to the lifeboat, which was in danger of being smashed against the side of the grounded vessel, two of the men belonging to the barque fell between the ship's side and the lifeboat. It was a terrifying moment, but they were hauled into the lifeboat uninjured.

Thirty-four people spent the night huddled together in the lifeboat, for the weather became thick and the lifeboatmen could not find the smacks that were standing by. Some of the lifeboatmen had remained in their boat for over 24 hours, with only a few dry biscuits to eat, and it was to be almost as long again before they returned to harbour.

When at daybreak the lifeboat at last came upon one of the smacks it was discovered that the steward had been left alone on the stranded ship. After the shipwrecked crew had been transferred to the smack the lifeboat went back again, but the seas were running 'mountains high', breaking over the deck amidships. Coxswain Thomas Tye took his boat under the jibboom, and the steward lowered himself into her down a rope.

Even when the tug *Harwich* took the rescued crew aboard and towed the *Springwell* into harbour the lifeboatmen's work was not finished. They went out at 5 a.m. the next day and helped the tug to haul the *Pasithea* off the sands and bring her into Harwich.

The first *Springwell*, a 35 ft. self-righting boat, proved rather too small for the work she was called on to do. It was impossible to launch her because of the surf on the beach during a vicious south-easterly gale on January 18th 1881, when a large barque was seen ashore on the Ridge, and a lot of time was wasted taking her on her carriage through the streets to the New Pier so that she could be lowered into the water by crane. This inability to launch seems surprising when one remembers that the lifeboathouse, which still stands, faces the harbour, not the open sea.

Soon after setting off, while still in the harbour, the lifeboat was struck by a heavy gust and capsized. She righted herself in about half a minute, but most of the crew were thrown out, and although eleven were picked up by boats from H.M.S. *Penelope* and the railway steamer *Claud Hamilton*, one was drowned.

Later in the year the R.N.L.I. sent a larger boat, also named *Springwell* and like her predecessor given by Miss Ellen Burmester, of London.

The first boat had saved the lives of 61 people, including the 23 taken off the *Pasithea*, and the second boat went on to save a further 82 lives.

But the lifeboatmen were not always welcome. Many skippers no doubt suspected they were only after salvage and regarded them as little better than pirates, remembering tales of the Swin Rangers. When Harwich lifeboatmen went to the assistance of the 305-ton Liverpool barquentine *Eboe*, owned by the African Association Ltd., on February 19th 1894, they found both the captain and the mate the worse for drink. The *Eboe* had struck the Longsand about low water, and after the anchor had been let go the captain decided he would get his ship off easily enough as the tide flowed. But the men either could not or would not obey his orders to heave up the anchor.

When the crew wanted to take to the boats, which had already been swung out, he loaded his revolver and threatened to shoot the first man to get into a boat. The barque was eventually refloated and taken into Harwich, but only with assistance.

Watermen and lifeboatmen crowded into the courtroom at Harwich a few days later when an English seamen and four Dutchmen were accused by the captain of refusing to obey orders. Several of the lifeboat crew gave evidence that the captain and mate were drunk, and one of them told the magistrates that when the crew were ordered to heave up the anchor the seas were so heavy that 'the poor fellows couldn't do it; the crew said if the lifeboatmen left the ship they would leave too —and very wise of them to do so'.

Heads nodded wisely in agreement, and a few minutes later the verdict of not guilty was greeted with loud applause from the waterside fraternity at the back of the court.

Plans for steam lifeboats had been mooted on many occasions, but until 1888 the difficulties inseparable from the use of steam power in a lifeboat were too great to be overcome to the satisfaction of the R.N.L.I. It was considered better to employ tugs to tow the pulling and sailing lifeboats which were in general use than to fit engines to lifeboats, even though the lifeboatmen's difficulties could be almost insuperable when no tugs were available.

The Harwich lifeboatmen found themselves up against it one November day in 1884, when they manned the *Springwell* in answer to rocket signals but were unable to get out of harbour. Eventually the Great Eastern Railway steamer towed the boat out when she left for the Continent, but the passage out behind a fast steamer intent on keeping to her schedule was, in the lifeboatmen's own words, 'something dreadful'. Towing behind such a vessel proved to be more than uncomfortable, the seas breaking over the boat in sheets and drenching the crew.

Nor was this the only unsatisfactory aspect of dependence on tugs. Just how much a lifeboat depended on a tug in the days of sail, and how necessary it was that the tug owners and skippers should know and accomplish what was required of them was illustrated in 1881, when a fishing smack brought news to Harwich of a vessel on the Longsand at the same time that a carrier pigeon brought a message from the Sunk lightship of a steamer ashore on the Sunk sand.

The *Springwell* started at noon, towed by the paddle-tug *Harwich*, reaching the Sunk two hours later. The steamer had foundered and the crew had been rescued by a passing vessel, the lifeboatmen discovered.

Tugmen and lifeboatmen then turned their attention to the vessel on the Longsand, about three miles away. But before reaching the wreck—the Liverpool barque *Indian Chief*—the owner of the tug, who was on board, decided that there was too much risk and, without consulting the lifeboatmen, ordered the skipper to turn for home. The lifeboat coxswain felt that nothing could be done without the tug and so the lifeboat returned to Harwich, reaching harbour about 5 p.m., by which time the *Indian Chief* had been ashore for more than fourteen hours. Some hours afterwards the Great Eastern Railway steamer towed the lifeboat out, but the boat was practically dragged through the water in the wake of the big packet, the seas sweeping over the boat and half-drowning the crew until eventually the towrope broke.

On this occasion the Ramsgate lifeboat *Bradford* was better served by the tug *Vulcan*, but she had much further to go than the Harwich boat and by the time she had reached the Longsand it was dark and the Ramsgate men could not find the wreck. Coxswain Fish and his men then took a most courageous decision to stay by the sands until daybreak.

There was no tug to take the Clacton lifeboat *Albert Edward* to the Longsand, but she arrived about 9 p.m. The crew found the wreck but discovered to their intense disappointment that she was the other side of the sandbank and they could not reach her. They anchored to wait for high water, but about two in the morning the light on the wreck disappeared, leading the lifeboatmen to believe the masts had gone overboard. They weighed anchor in a hurry and sailed up and down trying to find the ship in the darkness, but after about two hours decided to anchor again until daybreak. Then in the light of day they sailed up to the wreck, but could see no sign of life.

The Ramsgate boat had already managed with great difficulty to take off the 12 survivors. The other 17 members of the crew had been washed overboard during the night when the main and mizzen masts were carried away. One cannot escape the feeling that if the tug's owner had been willing to take a risk the previous day all 29 might have been saved.

In 1889 the R.N.L.I. had its first steam lifeboat, the *Duke of Northumberland*, built by R. & H. Green at Blackwall, on the Thames, and the following year she was sent to Harwich. With her arrival the *Springwell* became Harwich No. 1 and the steamer was known as Harwich No. 2.

The *Duke of Northumberland* was an interesting craft. In contrast to other British lifeboats she was built of steel, and in an attempt to avoid the danger of fouling a screw when working alongside a wreck or in shallow water the builders turned to hydraulic propulsion. Water was pumped in through openings in the bottom and out at the side—in fact she was the forerunner of the 'jet-propelled' speedboats of recent years.

Her first service was on October 8th 1890, when with the reserve lifeboat which had temporarily taken the place of the second *Springwell* at Harwich she stood by the brigantine *Ada*, of Faversham, a 172-ton vessel built at Prince Edward Island in 1872. In the same month she towed the sailing lifeboat out to the Shipwash and then assisted the tug *Harwich* to tow the 913-ton steamer *Achilles* off the sand after the lifeboatmen had helped to throw overboard some of her deck cargo of railway sleepers.

The *Duke of Northumberland* again proved her worth on November 13th 1890, when she and the reserve lifeboat helped to salvage the schooner *Christine Elizabeth*, of Haugesund, from the Longsand. The schooner was taken into Harwich for repairs and it was not until more than two months later that she was able to resume her voyage to the East Indies after being hauled out on the Naval Yard, where she was recaulked, recoppered and given a new mainmast.

Snow-clouds filled the sky when the little billy-boy ketch *Days* left London on Christmas Eve 1890, with a cargo of scrap-iron for Sunderland. On board were her owner-skipper, Henry Sweeting, the mate, Edward Hamilton, and a 17-year-old boy, George Lewington, who had bade his parents farewell and left his home at Grays not long before.

She sheltered from the weather for several days off Southend,

The wreck of the German emigrant ship Deutschland *on the Kentish Knock in 1875. The paddle tug* Liverpool, *seen on the extreme right, saved 155 but more than sixty lives were lost.*

An interesting view of Harwich harbour in the 1890s. The first City of Glasgow *steam lifeboat is at moorings and alongside Halfpenny Pier lies one of the local tugs, probably the* Spray*; just entering the picture is one of the Great Eastern Railway paddle steamers plying between Ipswich and Harwich; and at anchor out in the harbour are the guardship, H.M.S. Mersey, and a torpedo gunboat.* *(Photo: Mrs. S. H. Cooper)*

leaving there on January 5th 1891, to make her way slowly along the Essex coast against the bitter, snow-laden wind. Midnight found her off Harwich, and in the early morning, struggling through a blinding snowstorm, she struck on the Cork Sand.

The skipper rushed forward with the leadline. Five fathoms. . . He cast the lead again. Eight fathoms. . .

The worst is over, he thought. But a minute or two later there was a crash as she struck heavily on another part of the sand, and when the skipper dashed below to get a light he found himself up to the waist in water. He and his two companions tried to launch their small boat, but the waves caught it and smashed it against the billy-boy's side and stove it in. As the wreckage of the little boat was carried away by the swirling waves the ketch sank beneath them. All three climbed into the port rigging, the skipper lashing the youngster near him so that he could lean over and try to warm him by rubbing his face and hands.

'George, keep up lad, the lifeboat's coming,' the skipper told him. But no lifeboat came, and after they had clung to the rigging for five long agonising hours the boy died. It was not until dawn that the light-keepers aboard the Cork lightship spotted the wreck and signalled for the lifeboats.

The steam lifeboat left Harwich about 9 a.m., seven hours after the ketch had struck, with the pulling-and-sailing lifeboat in tow. When, nearly two hours later, the lifeboatmen saw the masts of the *Days* above the grey water the steam lifeboat swung round so that the men in the sailing lifeboat could attempt to make their grapnels fast in the rigging.

Time and again the grapnels broke away, but at last a line was thrown into the rigging and first one and then the other shipwrecked sailor was hauled through the surf to the lifeboat. Coxswain William Tyrell, of the *Duke of Northumberland*, looked at the two rescued men and then at the boy's body still hanging in the rigging. But he must not wait to recover the boy for he was already dead, and his business was with the living. He swung the wheel and set course for Harwich, where the two rescued men were cared for at the Great Eastern Hotel.

Later in the day the steam lifeboat returned to the wreck to recover the body of the boy. One of the crew, John Mills, jumped from the *Duke of Northumberland* into the rigging to cut the lashing which held the corpse to the shrouds—then he had to be dragged through the water on the end of a rope just as Skipper Sweeting and Edward Hamilton had been a few hours earlier.

Those were busy days for the Harwich lifeboats. Ears were sharply tuned in those days, and anything that sounded like a signal for the lifeboat brought crowds dashing down to the harbour. Ever on the alert, the Harwich lifeboatmen assembled on one occasion when a new steamer running between London and Harwich came into harbour to be welcomed by a gun salute which those hardy gentlemen mistook for the lifeboat signal. Another time some late-night revellers fired off a gun under the bedroom of the steam lifeboat's chief engineer, and those members of the crew within earshot were most annoyed when they discovered the joke that had been played on them.

By April 1891, the *Duke of Northumberland* had made 13 service trips, helping to save nearly 50 lives, as well as making ten trial trips.

The R.N.L.I. cannot have been dissatisfied with the use made of their new steam lifeboat at Harwich, but they wanted to gain further experience with the new type. There were other stations besides Harwich at which such a craft might prove useful, in spite of the limitations. So in 1892 the *Duke of Northumberland* was sent to New Brighton, being inspected by the Prince of Wales and the German Emperor when she called at Cowes on her way there. After five years at New Brighton the *Duke of Northumberland* was transferred to Holyhead and a new steam lifeboat was built to take over the New Brighton station.

In 1894 the Institution's second steam lifeboat, the *City of Glasgow*, was sent to Harwich. She was something of an improvement on the earlier boat, although having the same method of propulsion as the *Duke of Northumberland.* In place of the *Duke's* Thornycroft engine driving a single horizontally placed turbine, the *City of Glasgow* had an engine of Penn's design driving two vertical turbines to pump the water through the propulsion ducts. Like the first boat, the new one was built by R. & H. Green, who also built another steam lifeboat, similar to the *City of Glasgow*, for an Australian station.

The Harwich lifeboats were a focal point of interest, and officials of the South Holland Lifeboat Society inspected the Harwich boat before ordering a steam lifeboat from Thornycroft with machinery similar to that of the *Duke of Northumberland.* German lifeboat officials also included Harwich in their itinerary when visiting British lifeboat stations.

The *City of Glasgow* was transferred to Gorleston in 1897, but she performed only one service there before returning to Harwich early the following year.

Surf crashed on the beach, the southerly gale whipping up the spray and carrying it into the faces of the watchers ashore, as the Ipswich schooner *Rose* struggled along off Felixstowe on March 30th 1901. Her crew worked desperately to keep the vessel off the lee shore, praying that the tug *Merrimac* would reach them from Harwich in time.

Despite their efforts she blew ashore on St. Andrew's bank, near the Felixstowe coastguard station, only ten minutes before the tug arrived. Terrific seas swept across her decks as the 73-year-old vessel swung round, shuddering as her keel ground into the hard sand and shingle. Hundreds of people gathered along the shore to watch the coastguards launch their galley into the teeth of the gale, only to see it flung back into the breakers and capsized. Spectators rushed down to drag the coastguardsmen out of the water and to haul their boat up the beach as the steam lifeboat came in sight on her way from Harwich.

Several times the lifeboat tried in vain to approach the wreck, but each time the heavy seas flung her away. At last the lifeboatmen got a line aboard the *Rose* and brought their boat alongside. One by one, three sailors jumped from the wreck to the *City of Glasgow*. The fourth member of the crew leapt after them, but slipped and fell between the lifeboat and the wreck. Two lifeboatmen reached out quick as thought and caught his arm, hauling him on board.

The *City of Glasgow*, in spite of the good work she carried out, did not come up to expectations, and she was sold out of the service in 1901, when she was superseded by a second *City of Glasgow* built by J. Samuel White and Co. Ltd. at Cowes. The hydraulic system was abandoned in favour of a normal screw working in a tunnel, similar to that of a modern lifeboat.

The second *Springwell* was scrapped in 1902 and for a couple of years the steam lifeboat was alone at Harwich. Then in 1904 the Institution had another pulling and sailing boat, the *Ann Fawcett*, a 43 ft. Watson type, built by Thames Iron Works for the Harwich station. She did not do very much work at Harwich, for the only rescues with which she is credited were carried out in 1911 and 1912, most of the services being carried out by the steam lifeboat. In 1912 the Harwich No. 1 station was closed down and the *Ann Fawcett* went to Kingstown (Ireland) the following year, being renamed *Dunleary*. She remained there until 1919.

At this period Harwich was being used as a trial station for motor lifeboats, a number of which were put through their paces in the harbour and off the Essex coast before going to their stations. In 1907 the former Ramsgate pulling and sailing lifeboat *Bradford* spent nearly six months at Harwich after being converted to motor power, later going to Seaton Snook, and in 1909 the motor lifeboat *Charterhouse*, built by Thames Iron Works for Fishguard, spent ten days undergoing trials at Harwich. Another motor boat by the same builders, the Seaham lifeboat *Eliot Galer*, was tested at Harwich in 1911 and the following year two lifeboats for Scottish stations, the *William Macpherson* for Campbeltown and the *Alexander Tulloch* for Peterhead, were at Harwich on trials. The Harwich crew had the honour in 1913 of taking King George V for a trip in a new motor lifeboat they were testing.

One of the busiest days in the whole history of the Harwich lifeboats came on April 5th 1911. The first call that day was to the s.s. *Humber*, which had been set off course and was rolling heavily in Dovercourt Bay. She got into harbour without assistance, however. On her second trip the *City of Glasgow* towed the Hull brigantine *Volant* into Brightlingsea, and while she was away a third call was received. This was answered by the sailing lifeboat *Ann Fawcett*, which put ten men aboard the Faversham schooner *Lenore* and assisted her into harbour.

The Walton motor lifeboat *James Stevens No. 14* also launched to assist the *Volant*. As she set off one of the crew, James Spurgeon, was carried overboard but was rescued by his friends and remained in the boat while she sailed out to meet the *Volant*.

The *Ann Fawcett* had another trip in June of the same year, when she helped to refloat the Dutch schooner *L'Espoir de L'Avenir*, aground off Shingle Street, on the Suffolk coast. The skipper, his wife and daughter and the crew of three were preparing to leave when the lifeboat arrived, but Coxswain Ben Dale put three of his lifeboatmen aboard and sailed back to fetch a tug, which eventually hauled the schooner off.

The *City of Glasgow* carried out some valuable wartime services, although her usefulness was rather curtailed by the Naval regulations in force at the time.

She spent 30 hours at sea on November 21st and 22nd 1914, after putting out in response to urgent messages from the Walton lifeboatmen, who were standing by a Danish steamer aground on the West Sunk. In wintry weather the lifeboatmen set to work to jettison some 480 tons of her 2,000-ton cargo of coke and thus succeeded in refloating the ship.

Earlier in 1914 there had been an exciting race between the lifeboat and a tug after the lifeboat had put out in the teeth of a gale to assist the little ketch *Malvoisin*, of London, which was in distress off the Deben. The ketch was bound from Lynn to Rainham Creek with potatoes when her sails carried away. The tug won the race, but the lifeboat was the first to get a line aboard the disabled ketch—and the coxswain promptly engaged the services of the tug to help get her into harbour.

When the Norwegian steamer *Obidense* stranded on the Shipwash on the first day of 1915 her crew of 44, with the exception of the master and mate, took to the boats and were picked up by naval vessels and a collier. The *City of Glasgow*, which had left harbour to render assistance, returned on hearing that the crew had been picked up but later put out again to rescue the master and the mate. She got alongside after many attempts, but the lifeboatmen could not make fast as heavy seas were striking the wreck and washing completely over the lifeboatmen time after time. It took nearly an hour to get the captain and mate aboard. Then the lifeboat picked up another 25 members of the crew from the collier they had boarded and took them ashore.

In September the same year the lifeboat landed survivors from the mined Dutch passenger vessel *Koningin Emma*, whilst the last service recorded on the Harwich station's service boards was to another Dutch vessel, the s.s. *Zeeland*, in January 1916, when she assisted to save the ship.

The *City of Glasgow* was taken over by the Admiralty at the end of 1917, being renamed *Partick* and employed by the Navy on patrol duties. When the Navy eventually released her she left the East Coast, and is said to have found her way to the Nile. Thus the Harwich station closed down.

For fifty years the area covered for so long by the Harwich lifeboats was looked after by the Aldeburgh and Walton boats, but in 1967 the R.N.L.I. decided to re-open the Harwich station for an experimental period of two years. Once again Harwich was in the forefront of development, for the decision to re-open the station was prompted by the need to gain experience with the new 44-feet high-speed lifeboats being built by Brooke Marine Ltd. at Lowestoft to a United States Coast Guard design.

The new lifeboats, built of steel like the steam lifeboats of sixty years before but of mainly welded construction instead of rivetted, are twin-screw boats of a radically different design from anything employed in the British lifeboat fleet in earlier years. Instead of the eight knots of a normal motor lifeboat they are capable of a speed of 14 knots for short periods and have an operational range of 205 miles at 12 knots or 305 miles at 10 knots. The only lifeboat with such a speed as this until the arrival of the American design was the Dover lifeboat *Sir Richard Hillary*, which was intended primarily for dealing with aircraft casualties in the Straits of Dover.

The *Margaret Graham* was named at Harwich on September 27th 1967, the fourth of the six steel lifeboats built by Brooke Marine. Instead of the normal crew of eight or nine the American-type lifeboats carry only five, headed by a full-time coxswain-mechanic. The crew members have to be much more than just excellent seamen; they must know how to operate the electronic equipment, including radar, which is fitted.

Though the *Margaret Graham*'s first call proved to be a false alarm she carried out a useful service on October 17th 1967, when she battled her way against winds of force nine, gusting to force ten, to reach the Norwegian tanker *Astrid Elizabeth* which had a leak in the engine room about 22 miles from the harbour, in the vicinity of the Sunk. It took about 2½ hours to reach the casualty, which the lifeboatmen found at anchor. The lifeboat stood by while the German tug *Heros* arrived from Dover and took the tanker in tow, and then returned to harbour.

Next day she was out again to the same area after a report had been received of a catamaran drifting five miles from the Sunk lightship. This was a vessel from which Clacton lifeboat had taken a number of men the previous day.

So a new chapter in the story of the Harwich lifeboats has begun and there is no doubt the work of these boats will continue for many years.

BOAT RECORD, HARWICH

Years on Station	Length, Breadth, Oars/Crew	Type, Weight, Cost	Year built, (Official No.), Builder	Boat's name, Donor, Authority
1821-1825+		N&S?	1821 George Graham, Harwich	*Braybrooke* Essex L.B. Association
1876-1881	35′ 9′ 10 db	SR £432	1876	*Springwell* Miss E. Burmester, London R.N.L.I.

1881- 1902	45′ 11′ 12 db/15	SR £600	1881 (317) Woolfe, Shadwell	*Springwell* Miss E. Burmester, London R.N.L.I.
1904- 1912	43′ 12′ 6″ 10/13	W 11t 14	1904 (517) Thames Iron Works, Blackwall	*Ann Fawcett*[1] leg. Mrs. Fawcett, London R.N.L.I.
1890- 1892	50′ 12′	Steam	1889 (231) R & H. Green, Blackwall	*Duke of Northumberland*[2] R.N.L.I. funds R.N.L.I.
1894- 1901	52′ 15′ 2″	Steam	1894 (362) R. & H. Green, Blackwall	*City of Glasgow*[3] Glasgow LB Fund R.N.L.I.
1901- 1917	56′ 14′ 8″	Steam 32t dis	1901 (446) J. S. White, E. Cowes	*City of Glasgow*[4] Glasgow LB Fund R.N.L.I.
1967-	44′ 12′ -/5	U.S.C.G. 17t 16 £40,000	1967 (1004) Brooke Marine Ltd., Lowestoft	*Margaret Graham* Anon R.N.L.I.

[1]Transferred to Kingstown, Co. Dublin
[2]Transferred to Holyhead
[3]At Gorleston for short time, 1897-1898
[4]Taken over by Admiralty, 1917

SERVICE RECORD, HARWICH No. 1

Springwell Lifeboat

1876	Dec.	12	Schooner *Exhibition*, of Colchester, rendered assistance	
	Dec.	29	Barquentine *Willy*, of Llanelli, rendered assistance	
1877	Dec.	2	Barque *Jacob Lanstrum*, of Gothenburg, saved	8
1879	Feb.	16	Barque *Pasithea*, of Liverpool, saved	23
1881	Jan.	6	Spanish schooner *Rosita*, saved	11
	Jan.	20	Dutch steamer *Ingerid*, saved	7
	Oct.	23	Barque *Iris*, of Gefle, saved	12
	Oct.	24	Barque *Iris*, of Gefle, rendered assistance	

Springwell (second) Lifeboat

1881	Dec.	17	Barque *Olive Branch*, of Harwich, saved	3
1882	Apr.	25	Schooner *Henrietta*, of London, stood by vessel and rescued	4
1883	Feb.	7	Barque *Loreley*, of Memel, saved	11
1884	Jan.	24	Barge *Jessie*, of Rochester, saved vessel	
	Oct.	11	Barque *Morford and Trubey*, of Aberdeen, saved	3
1886	Dec.	4	Ship *Constanze*, of Hamburg, assisted to save vessel	
1888	Mar.	4	Brig *William and Anthony*, of Folkestone, saved vessel and	8
	Nov.	9	Brig *Carl Gustav*, of Christianstad, saved	9
1890	Mar.	2	Schooner *Mary and Maria*, of Hartlepool, assisted to save vessel and	3

A Temporary Lifeboat

1890	Oct.	8	Brigantine *Ada*, of Faversham, stood by vessel	
	Oct.	20	s.s. *Achilles*, of Sunderland, assisted to save vessel and	21
	Nov.	13	Schooner *Christine Elizabeth*, of Haugesund, assisted to save vessel	
1891	Jan.	6	Ketch *Days*, of Barrow, saved	2

Springwell (second) Lifeboat

1891	Nov.	11	Smack *Brunette*, of Colchester, rendered assistance	
1892	June	5	Barque *Ephrussi*, of Brevig, stood by vessel	
1893	Jan.	17	s.s. *Helsingor*, of Elsinore, landed 18 from Longsand l. v.	
	Nov.	4	s.s. *Rockcliff*, of West Hartlepool, rendered assistance	
	Nov.	10	Barque *St. Olaf*, of Mariehamn, saved	12
1894	Feb.	19	Barque *Eboe*, of Liverpool, assisted to save vessel and	13
1897	Jan.	23	Schooner *Sancho Panza*, of Faversham, saved	6
	Jan.	23	Fishing smack's boat, saved	5
1898	Oct.	17	Barque *Inga*, of Laurvig, stood by vessel	
1899	Nov.	2	Barque *Pactolus*, of London, saved	5

Ann Fawcett Lifeboat

1911	Apr.	5	Barquentine *Lenore*, of Faversham, saved vessel and	6
	June	25	Barque *L'Espoire de L'Avenir*, of Rotterdam, assisted to save vessel and	6
1912	Jan.	6	Barge *Monarch*, of London, assisted to save vessel and	3

HARWICH No. 2

Duke of Northumberland Steam Lifeboat

1890	Oct.	8	Brigantine *Ada*, of Faversham, stood by	
	Oct.	20	s.s. *Achilles*, of Sunderland, assisted to save vessel and	21
	Nov.	13	Schooner *Christine Elizabeth*, of Haugesund, assisted to save vessel	
1891	Jan.	6	Ketch *Days*, towed No. 1 boat to ketch	
	Mar.	3	Schooner *Mercury*, of Aberdeen, saved	12

City of Glasgow Steam Lifeboat

1895	June	6	Schooner *Hans*, of Rendsburg, assisted to save vessel and	4
1897	Oct.	25	Barque *Triton*, of Rostock, assisted to save vessel	
1898	Mar.	26	Ship *Pampa*, of Hamburg, assisted to save vessel and	24
1899	Jan.	30	Longsand lightvessel, landed 2	
	Dec.	7	s.s. *Lambeth*, of London, landed 8 from Sunk lightvessel	
1901	Mar.	30	Schooner *Rose*, of Ipswich, saved	4

City of Glasgow (second) Steam Lifeboat

1901	May	17	Schooner *Harriet*, of Goole, stood by vessel	
1904	Nov.	23	s.s. *Tyne*, of Newcastle, assisted to save vessel	
1905	Jan.	21	Schooner *Dashing Wave*, of Fowey, assisted to save vessel	
	Nov.	23	Barquentine *Thirza*, of Faversham, assisted to save vessel and	7
	Nov.	23	Tug *Spray*, of Harwich, rendered assistance	
1906	Dec.	25	Barque *Earlshall*, of Dundee, stood by vessel	
1907	June	19	s.s. *Ariadne*, of London, stood by vessel	
1908	Apr.	7	Schooner *Notre Dame des Toutes Aides*, of Nantes, saved	9
	Sept.	9	Barge *San Pedro*, of Rio de Janeiro, saved vessel and	5
1909	May	14	Smack *Tripper*, of Harwich, stood by vessel	
1910	Aug.	19	Barque *Fox*, of Arendal, stood by vessel	
	Oct.	14	s.s. *Hurricane*, of Liverpool, stood by vessel	
	Oct.	27	s.s. *Baltzar von Platen*, of Helsingborg, rendered assistance	
	Dec.	16	Barge *Baltic*, of London, stood by vessel	
1911	Apr.	5	Barquentine *Volant*, of Harwich, assisted to save vessel and	6
	Oct.	30	Barge *Antje*, of London, saved	3
1912	Dec.	30	Barge *Agnes and Constance*, of Rochester, saved vessel and	3
1913	Aug.	21	Schooner *Christable*, of Whitstable, assisted to save vessel	
	Oct.	25	Trinity House pilot cutter *Will o' the Wisp*, saved boat and	9
1914	Mar.	18	Ketch *Malvoisin*, of London, assisted to save vessel and	5
	Nov.	21	s.s. *Lilian Drost*, of Copenhagen, assisted to save vessel and	18
1915	Jan.	1	s.s. *Obidense*, of Bergen, landed 25, saved	2
	Sept.	22	s.s. *Koningin Emma*, of Amsterdam, saved	20
1916	Jan.	7-8	s.s. *Zeeland*, of Rotterdam, assisted to save vessel.	

3

WALTON

WALTON-ON-NAZE lies some six or seven miles to the southward of Harwich, more or less under the shadow of the Naze, a high-standing promontory on which stands a lofty column of brick, the Naze Tower, built in the eighteenth century as a seamark. Westward of the town, behind the Naze, lie the Backwaters, an intricate series of channels known well enough to the skippers of spritsail barges trading to the various small quays which lay up the Backwaters but little known to others.

There was a busy little fishing fraternity at Walton in the latter part of last century whose members went a'pleasuring in summer and fishing in winter, and fitted in a bit of salvaging when opportunity offered. They had a number of big boats which were employed from time to time on salvage work even after the coming of the lifeboat; when the s.s. *Brighton* went on the Gunfleet in fog in 1892 and the Harwich and Walton lifeboats went to her assistance 'a private boat' was rowed out to their aid from Walton by members of the fishing community, and a couple of years later the Polleys, the Oxleys and Fred Sparrow went out in the large boat *Who'd 'a Thought It* and did some 'salvaging' in the case of the brigantine *John White*, of Faversham, which had been run on the Gunfleet after springing a leak.

At this period there were some very big pleasure boats belonging to the watermen at Walton. Biggest were the *Skylark*, which could carry 97 passengers, and the *White Swan*, which was licensed for 96.

A lifeboat had been mooted at Walton for some years before the R.N.L.I. allocated one to the town in 1884, and in fact it does not seem that this was the first lifeboat the place had known. Somewhere about 1880 a visitor to the town had suggested that a lifeboat should be provided and several hundred pounds had been collected for the purpose, but this scheme did not materialise and the money went elsewhere. Whatever happened to that project, a feature of Walton watermen's regatta in 1880 was the christening and trial of a private lifeboat belonging to George Polley, a prominent local waterman.

The R.N.L.I. boat supplied to the station in 1884 was a 37 ft. self-righter built at Limehouse by Forrestt & Son and paid for by the drama group of the Honourable Artillery Company, one of the most famous of English regiments, which had by chance been stationed at Walton about 1860. The old lifeboat house, at the north of the sea front just beyond the Coastguard station, still bears the plaque recording that the H.A.C. were the donors of that first boat. The regiment provided a guard of honour when she was named on November 18th 1884.

The crew was at a celebration dinner at the Queen's Head Inn the evening after the naming ceremony when Captain Costick, the chief officer of the local Coastguards, was heard shouting 'Come at once, men, you're wanted—signals of distress going up!' Lifeboatmen have always been ready for service at a moment's notice, but the sight of a lifeboat going off with one of the crew resplendent in Yeomanry uniform and the rest in 'Sunday rig' can hardly be a common one.

Unhappily, it was an inauspicious start. Not only were several best suits ruined, but the slipway proved too narrow, there was no room to guide the carriage and the lifeboat was launched precipitately at the foot of the slipway, the crew being thrown against one another into the bottom of the boat. Then, when they returned after an all-night search in which nothing was found, the boat was dashed against the bottom of the slipway and stove in.

The crew had little time to brood on this, however, for the *Honourable Artillery Company* made her first rescue on Boxing Day 1884, when she took 25 men from the German full-rigged ship *Deike Rickmers* (2,600 tons), belonging to T. J. Rickmers, of Bremerhaven. 'Ship requires immediate assistance' was acknowledged by Walton Coastguards shortly after noon on Christmas Day, but nobody knew the whereabouts of the vessel in trouble and it was not until both the Gunfleet lighthouse and the Sunk lightship sent up rockets at sunset that the chief officer decided to send the lifeboat off—and then because of the unsuitable slipway it was another hour and a half before the lifeboat could be got from the boathouse on to the slip.

Crossing the broken water on the Gunfleet, the lifeboat beat down Swin against a strong nor'easter to the Sunk lightvessel, whose crew directed them to the Longsand. They brought up to windward of the wreck and waited for daylight. The sky was hardly beginning to pale, when, at 6 a.m., the men on the wreck fired a rocket, and seeing this the lifeboatmen neared the ship, let go the anchor and manoeuvred their craft under the ship's jibboom.

Forty feet above them the lifeboatmen could see faces. Some of the crew were on the bowsprit, others peered over the bows. Coxswain Henry Britton asked how many there were.

'Twenty-five', shouted Captain Reents. 'I can take you all,' shouted back the coxswain. Captain Reents was doubtful, and wanted to wait until another boat arrived, but Coxswain Britton knew what he was about.

'I shan't leave until I have all of you,' he yelled above the roar of the wind. First to leave was the crew's pet dog, which was lowered into the lifeboat on the end of a line. Then the crew slid one by one down a rope from the end of the jibboom, and finally only the captain was left. It was a tragic end to Captain Reents' first voyage in the ten-year-old *Deike Rickmers*—a voyage which had begun at Bremerhaven only two days before—and he hesitated to leave.

But the coxswain was getting anxious and the crew shouted to their captain to leave her in time. When he did drop into the lifeboat there were 40 men—and the dog—in the little boat, and the crew had to work under great difficulties. Unable to get round the Longsand Head against the flood tide, they had to wait four hours for sufficient water to cross the Longsand, and it was 7 p.m. before they arrived home.

The German seamen were taken to Dorling's Hotel, there to have a Christmas dinner of roast beef, goose and plum pudding, hardly the sort of meal they could have hoped for twelve hours before. When, after the crew had had their meal, they began singing some of their German songs the Walton men responded with the old revivalist hymn 'Pull for the shore, sailor,' surely a most appropriate choice.

The inadequate slipway already mentioned proved a thorn in the side of the Walton lifeboatmen for quite a long time. Tests showed that even in good conditions it took 25 minutes to launch the boat, and alterations to the slip never really made it suitable for its purpose. Then in 1892 the slipway claimed the life of one of the crew.

The *Dilsberg*, a steamer of 967 net tons built at Port Glasgow in 1882, went on the Kentish Knock the day after leaving Middlesbrough with a cargo of pig-iron for Civita Vecchia. She bumped so heavily that the crew could hear the stanchions breaking. To add to the vessel's difficulties an engine-room accident in which the second engineer was badly scalded meant the crew could not get steam on the donkey engine to pump the ship out. They could see the Sunk, Kentish Knock and Longsand lightships, but nothing much was done by the men aboard them to attract assistance from the shore.

The old Cork lightvessel, a wooden lightship typical of those stationed off the East Coast in the early years of the century. It was the signal guns carried by these lightships that often gave the first warning of wrecks on the sandbanks—a code was used to indicate in what area the wreck had occurred—and it was to these lightvessels that the lifeboatmen often went for information when carrying out a search.

Coasting barges were often the cause of Essex lifeboats being called out. Here a ketch barge passes a spritty, with a deck cargo of timber, at anchor. Further out a big mulie—a spritsail barge with a large gaff mizzen—beats to windward.

By the time two of the boats had been made ready, heavy seas were washing over the *Dilsberg* from stem to stern, carrying away the wheel-house and smashing the hatches. Six of the crew were swept away, but 15 survivors, after spending a night on the crippled ship, managed to row to the Kentish Knock lightvessel in a damaged boat which they patched with a piece of tinplate. The captain died from exposure on the way to the lightship, from which the survivors were later taken by the Ramsgate smack *Cordelia.*

It was not until several hours after the crew had abandoned their ship that the signal calling the lifeboat crew was heard at Walton. There was a rush for the lifeboat house, two or three men going after each lifebelt, and one group decided to toss for a belt. Jovial John 'Darkie' Downes, who had already had 31 trips in the boat, lost the toss—but stayed to lend his weight on one of the launching ropes.

'Are you all ready?' asked Coxswain Britton. 'All ready.'

'Stand clear everybody,' called Mr. Pengelly, the Coastguard chief officer who took charge of the launching party. 'Let go!'

As the boat ran on its carriage down the slipway it swerved to the side. 'For God's sake get out of the way!' yelled one of the launchers, knocking two of his fellow launchers down to get them clear of the wheels. One man did not let go of the drag rope, and as it tightened he was flung under one of the carriage wheels. It was 'Darkie' Downes.

'I'm crushed, I'm a dead man,' he told one of his colleagues who lifted his head as he lay on the slipway. And within a few hours it was so.

The *Honourable Artillery Company* was eventually launched and was towed out to the Longsand by the paddle tug *Merrimac*, belonging to R. & W. Paul, the Ipswich maltsters. The lifeboat and the tug had to bring up until daylight, when the lifeboatmen were able to board the wrecked steamer at low water. The *Dilsberg's* rudder had gone, the decks were ripped up and the hatches burst open. The only man on board was a dead seaman whom they found in the forecastle.

Sometimes the weather off the Essex shore can change with startling suddenness. So it was in November 1893. On the 18th the sun was shining brightly and the sea as calm as a pond, but next day the wind was blowing half a gale from the nor'ard and the sea was tumbling over the Gunfleet. Watchers at Walton saw a three-masted schooner in ballast come through the Gatway to seek shelter off Holland, running like a racehorse with only the smallest of sails set; and they muttered among themselves as a ketch found it impossible to weather the Naze and had to run back.

The lifeboat was launched the following day when a barque that had been riding at anchor two or three miles above the Gunfleet lighthouse was seen to be ashore. She sailed out through the Gatway to find that the barque, the *Elphinstone*, of Christiania, had driven well up on the Heaps with the seas breaking about her unmercifully. Could anything live amid that, the lifeboatmen asked themselves?

The lifeboat was anchored and veered down towards the wreck, the crew pulling at the oars for nearly an hour in an effort to bring the boat alongside. The breakers swamped the lifeboat time after time and the relieving valves became choked with sand. One lifeboatman actually struck out swimming as the boat filled to the gunwales, thinking he had been washed overboard.

'I'd give a hundred pounds this moment if we could have saved the poor fellows,' muttered one of the Walton men as they stared at the wreck and saw nobody. But a moment later they caught sight of a man peering through the wind-driven spray from the deckhouse. He lifted his hands, called to others, and within a few moments was joined by several more men. Realisation that they were not too late after all spurred the lifeboat crew to yet greater efforts, and at last they managed to pull alongside the wreck.

'Come steady like—not all at once!' they shouted as five or six men tried to jump into the lifeboat at the same time. Fourteen men were in the boat, as Coxswain Britton said afterwards, 'in a jiffy before I could hardly realise it'.

In 1894, following a stormy disagreement among the crew of the Institution boat over whether the bowman was competent and fit for the job—and possibly other matters as well—several men left the crew and bought a second-hand lifeboat which they renamed *True-to-the-Core* and ran as a private lifeboat. For the next twenty years there was intense rivalry between the Institution lifeboat and the private one. It may be an exaggeration, but it is said that the rivalry was on occasion carried on ashore by the respective crews in a very spirited and rather violent manner.

This private lifeboat was a powerful craft of the Norfolk and Suffolk type, rigged with a dipping lug foresail and a standing lug mizzen, a very different type from the *Honourable Artillery Company*. It seems almost certain that this private lifeboat was built by Beeching at

Yarmouth in 1854 for the beachmen at Scratby, on the Norfolk coast. She was then named the *Royal Albert*.

The *True-to-the-Core* was quickly in action, for no sooner had she arrived off Walton on her delivery trip in May 1894, than she was diverted to the aid of the Harwich fishing boat *Who'd 'a Thought It*, which was in trouble in the Wallet. After a trial trip to Harwich and a display for visitors, the *True-to-the-Core* was out in earnest on August 24th when she and the *Honourable Artillery Company* carried out a lengthy but abortive search after signals had been heard from the lightships off the coast. Three days later the volunteer lifeboat went out to the barquentine *Britannia*, aground on the Gunfleet after mistaking the red light of the lighthouse for a vessel's navigation light. She stood by until the barquentine was towed off, but whether the services of a lifeboat were really needed is doubtful, as the pleasure boats *White Swan* and *Skylark* ran trips out to the stranded vessel during the afternoon to give summer visitors a close-up view of the salvage operations.

The introduction of the volunteer lifeboat was accompanied by some very unfortunate ill-feeling among the local watermen, apparently fanned into flame by visitors and local hotheads. 'Fanatical politicians have endeavoured to even make political capital by fostering the dispute,' the *Essex Standard* rather ungrammatically stated in September.

While the *Honourable Artillery Company*'s crew had to obey the rules of the R.N.L.I. and could only launch on receiving definite instructions, the crew of the *True-to-the-Core* were a law unto themselves and would often put off in search of salvage when there was in fact no call for lifesaving. This was a constant source of complaint on the part of the Institution crew, whose supporters claimed that the private lifeboat crew was nothing more than a salvage company, like the beach companies of Norfolk and Suffolk, and as such was not entitled to the donations which were solicited from holidaymakers and others at every opportunity. Such propaganda was, of course, hotly refuted by the champions of the volunteer crew.

The controversy even reached Parliamentary circles later in 1894 when E. H. Bayley, the local M.P., wrote to the First Lord of the Admiralty calling his attention to the Admiralty's refusal to allow messages received by the Coastguard from the Gunfleet lighthouse to be passed to the coxswain of the *True-to-the-Core*.

Before long a notice was posted at the Coastguard station saying that 'All messages received from lighthouses respecting wrecks or vessels in distress will be posted on this Board. When received, a flag, the Blue

Peter, will be hoisted in the daytime at the masthead; and at night a light in the same place to intimate that news has been received.' Thus was that question settled for the time being, but friction between the two crews continued unabated for some years and a similar question was raised in Parliament twelve years later.

The rivalry between the two crews was aggravated by the fact that the greater spread of canvas on the *True-to-the-Core* gave her a distinct advantage over the *Honourable Artillery Company* when hurrying to a wreck, particularly in light weather. This fact was very noticeable one December day in 1894 when both vessels put out in light winds, the *Honourable Artillery Company* attending to the disabled barque *Freia* and the private lifeboat sailing out to a brig stranded on the lower end of the Gunfleet, only to find that she had been abandoned.

There was also a race between the two lifeboats in January 1895, when several vessels were in difficulties in the Swin. This time the R.N.L.I. boat was first away, the ferrying of the crew to the private boat—which lay afloat near the pier—causing her to be eight minutes behind at the start. As there was something like a gale her bigger sail was not in use, but the men of the *True-to-the-Core* were not to be beaten, and lowered their small lugsail and set the big one, reefed. Watchers ashore saw that she was gaining on the self-righting boat as both lifeboats were swallowed up in a snow squall.

The *True-to-the-Core* took the crew of four off the 31-year-old schooner *Forest Oak*, a 75-tonner built at Sunderland in 1863, which had lost both her masts and had become unmanageable while on a voyage from Sunderland to London with a cargo of empty bottles. When they came ashore the crew of the private lifeboat were triumphant. 'Where's the ship?' they were asked by watchers on the pier. 'She's at the bottom by now,' replied Coxswain Haven Puxley, 'but we've saved the crew!'

Not so pleased were the men of the *Honourable Artillery Company* when they were questioned by the helpers who hauled their boat up the slipway. 'Their boat out-sailed and beat us—but only by a few minutes,' they admitted grudgingly.

The R.N.L.I. awarded 30s. to each of the 12 men making up the crew of the *True-to-the-Core* for the rescue of the *Forest Oak*'s crew, and the Board of Trade followed this with a grant of £2 to the coxswain and £1 to each of the men. First blood to the *True-to-the-Core*.

Little more than a week later news that a vessel was ashore on the Gunfleet sent both lifeboats out again. Henry Britton and others of the

The Walton lifeboat Honourable Artillery Company *outside the boathouse at Walton. (Photo: Frank Putman Ltd.)*

Above and below: The Walton volunteer lifeboat True-to-the-Core, *a Norfolk and Suffolk type boat built in 1897 to replace an old lifeboat bought from Norfolk three years earlier.* (*Photos: V. G. Scammell*)

Institution crew were 'down the river'—in the Walton Backwaters—with some gentlemen wildfowlers when the lifeboat signals went. The coxswain got ashore at once, landing on some very rough ground, but he had to cross saltings and a ploughed field and then run all the way down Hall Lane, and by the time he reached the lifeboat shed he was 'knocked up'. Then the lifeboat was delayed still further by an accident as she was being launched from the carriage, the rudder being damaged, and by the time she was afloat the *True-to-the-Core* had a good start.

The *Honourable Artillery Company* gained the lead, however, by crossing the sands where the deeper-draught boat could not go, and the two arrived at the casualty almost together. They found the Belfast brigantine *Opah* (113 registered tons), a product of the prolific Prince Edward Island shipyards, which had turned her out in 1858, stranded high on the sands, surrounded by water too shallow to allow the lifeboats to approach.

Two of each crew leapt overboard and waded waist-deep to the vessel. Nobody was aboard—the crew had got ashore in their own boat, landing on Clacton pier—and one man had to lift another on the fluke of the vessel's anchor so he could climb aboard and throw down a rope ladder for the rest to climb up. They waited for two hours until the tide had risen enough for the *Honourable Artillery Company* to get alongside; it was another three-quarters of an hour before the depth was sufficient for the *True-to-the-Core*.

Harry Britton realised his own crew could do little, so he waited for Haven Puxley and the volunteer crew to come aboard. Then both crews set to work to lighten the brigantine, but the cargo of chalk had frozen solid and there were not enough shovels. It was clear they could do no good, so they gave up and returned ashore. Sails and gear salvaged from the *Opah* were sold at Walton Coastguard station later, but the vessel herself seems to have left her bones on the sand.

There would seem to be some point in the criticism that the private lifeboat was nothing more than a salvage company, for the *True-to-the-Core* went out on occasions when the only object could have been salvage. Not every time was her crew successful. There was, for instance, the occasion in 1895 when they rowed all the way to the Barrow, more than ten miles, after a vessel had gone ashore there in a flat calm. The skipper refused to employ them, and they had to row all the way home again on a chilly, moonlight night.

The volunteer lifeboatmen were more successful later the same year, when the three-masted schooner *English Girl*, of Fleetwood, went on the Gunfleet. They went out and helped to jettison 40 tons of her cargo of firebricks and then assisted the *Merrimac* to tow her off the sand and into Harwich. That was a bad day for the Institution boat, which was called out soon after 1 a.m. There were few launchers available at that early hour, and as she was being hauled off her carriage into shallow water the boat ploughed her forefoot into the beach and all the launchers' efforts to free her were unavailing. With the tide ebbing fast the launch had to be abandoned.

One result of this fiasco was that some sour grapes were squeezed, and in the course of the renewed controversy it was alleged that the 15-man crew of the *Honourable Artillery Company* contained two bricklayer's labourers, one fitter, one turner and a foundryman. One the of R.N.L.I. men retorted that while he was employed as a bricklayer's labourer 'it answers my purpose to do that, or any other sort of labour, rather than lounging about against the hotel corners as some prefer to do. I am a bricklayer's labourer, but at the same time I am in possession of seaman's discharges for four years and six months.'

He retaliated with the information that the crew of the *True-to-the-Core* was made up of a fishmonger, a painter and paperhanger, an insurance agent, a Limehouse docker, a tobacconist, three bricklayer's labourers, a stage super, a baker, an oil merchant, two sons of clergy and one barman.

On September 12th 1896 both the Walton lifeboats went out to the assistance of the barquentine *Sunderland Packet*, of London, which had gone aground on the Gunfleet after missing stays in the Swin. Coxswain George Polley, who had taken charge of the *True-to-the-Core* when Haven Puxley resigned and joined the crew of the Institution boat, attempted to salvage the vessel, a 171-tonner built at Sunderland in 1856, but she had to be abandoned when the *Merrimac* nearly pulled her to pieces in an attempt to get her afloat. Wreckage from the vessel was later sold at Harwich.

The 164-ton Guernsey brigantine *Lord Strangford* was luckier some ten weeks later, being assisted off the Gunfleet by the *Merrimac* and the crew of the *Honourable Artillery Company*, which had been towed up Swin to the wreck by a steamer. There was no work for the men of the *True-to-the-Core* this time. The lifeboatmen had to jettison some of her cargo of scrap iron, throwing overboard numerous old bedsteads, broken bicycles, leaky baths and rusty pots and pans, and then as the tide flowed they had to set to work at the pumps, for the water rose inside as well as out. And at the pumps they had to stay as the *Lord Strangford*—an old vessel by any standards, for she had been built at Fishbourne on the Isle of Wight, reputedly as a yacht, in 1825—was towed into Harwich harbour, where she had to be run up on the mud. They were hard at work for 16 hours, so it is not surprising that they refreshed themselves at Harwich before sailing the lifeboat back to Walton against a strong wind and a heavy sea. Nor was the salvage of such an old vessel really worth all the effort, for the *Merrimac's* owners were awarded only £35 when the case came up for adjudication; what the lifeboatmen got was not recorded.

An old boat when she was acquired by the Walton men, the *True-to-the-Core* was not really fit for lifesaving work in heavy weather—"the fittings of her mizzen mast" gave way on one of her outings in November 1894, and she had to be towed into Harwich for repairs—so the Walton men did their utmost to raise money for a new boat. They received support from Colchester as well as from summer visitors to Walton, and in 1897 a new lifeboat very similar to the old one was built by John Houston at Rowhedge. Launched on October 12th 1897, she was 40 ft. long and cost £437.

The new *True-to-the-Core*, which lay at anchor off the pier and was reached by means of a boarding boat in much the same way as the present Institution lifeboat, was in danger of being wrecked herself in very heavy weather in the following March. This was a busy month for all the Essex lifeboats. On the 24th, Coxswain Schofield of Clacton received a telegraph message from Southend that vessels were aground on the Maplin and Barrow sands, and at noon the following day the Walton lifeboat was called to assist a vessel on the Barrow—the same vessel that the Clacton boat had been to earlier had run aground again it seemed. It was the most exciting launch seen at Walton up to that time; huge seas filled the lifeboat time and again before she could get away from the shore, and it took the most strenuous efforts of the crew and the launchers to keep her from being thrown on to the breakwater. The launchers' efforts did succeed in time, however, and the drenched crew set the sails just in time for the *Honourable Artillery Company* to weather the new pierhead. The seas washed the carriage off the slipway and the launchers had to salvage it at low water. Once in the open water the lifeboat felt the full force of the gale. She shipped sea after sea, and the crew had to lower the foresail and set the mizzen as a storm foresail, even this being almost too much sail for the conditions prevailing.

There was nothing to be seen on the Barrow when they got there; at the Swin Middle lightship the lifeboatmen were told, 'No, haven't seen anything since a fish trawler got off—but there's a large steamship up by the Maplin, we don't know what's the matter with her.' The trawler was the Hull Steam Fishing and Trading Company's *India*, a 194-gross ton vessel; the steamer near the Maplin was safely at anchor.

Finding nothing in need of their services the Walton men ran down to Sheerness, where they were the guests of the first-class petty officers' mess for two days. They found the Clacton crew already in the naval barracks as the guests of the chief petty officers.

When a call came for a lifeboat on the evening of March 26th it was the *True-to-the-Core* which responded. The *True-to-the-Core* and the Aldeburgh lifeboat—which had landed the crew of the West Hartlepool ketch *Ludovicus* at Harwich a couple of days before—ran like racehorses to the Maplin, where they found the 1,676-ton full-rigged ship *Pampa*, bound from her home port of Hamburg to San Francisco with 2,500 tons of cement, hard ashore. The Walton and Aldeburgh crews were joined by the men of the Harwich steam lifeboat *City of Glasgow* in jettisoning cargo; the three crews worked together all the next day and through the following night until high water at 3.30 a.m., when the tugs *Spray*, of Harwich, and *Cambria*, of London, towed the vessel off and took her to Gravesend.

On the way home the Aldeburgh lifeboat took the crew out of a big sailing ship on the east side of the Barrow sand. The Clacton lifeboat, which had arrived back from Sheerness on the 27th, was launched on the 28th to this vessel, the barque *Magdimenta*, of Kragero, which had a cargo of ice, by this time well mixed with seawater. The next day the Walton lifeboats were both out again, to a schooner high and dry on the Gunfleet with all sail set. The volunteer crew sent a small pulling boat across the Wallet 'to get first hold', but there was no salvage to be had; the vessel came off without assistance on the midday high water.

The fishermen and the men who manned the lifeboats knew the channels through the sands among which so much of their work was done as nobody else did, and from time to time they used their knowledge to prevent disaster. On occasion, however, their good advice was ignored, as for example in 1899 when the crew of the *Honourable Artillery Company*, returning from a call to the Galloper lightship after reports of a collision, fell in with a French schooner running up the Sunk channel. Knowing she was running into danger they followed and hailed her. 'Where bound?' 'Londres,' came the reply. The lifeboatmen warned the Frenchman he would run aground, but they were told in no uncertain manner—or language—to mind their own business.

Sure enough the schooner, the *Clementine*, of Nantes, went aground. The lifeboat lay by till daylight, in case the wind got up, though the French crew refused all offers of assistance. When the Harwich steam lifeboat came out the Walton men returned home, as they had no provisions.

The *True-to-the-Core* went out later, but the volunteer crew also were told to keep off. After some time, however, Fred Sparrow decided to make fast and go aboard. He told the French master he would have to lighten his ship to get her off. The master had learned his lesson; he told Fred Sparrow to go ahead, and the lifeboatmen went to work at the pumps. The tug *Spray* eventually towed the schooner off and she was taken into Harwich.

When the *True-to-the-Core* got back in the early hours of January 8th she had been out 42 hours, and in all that time the crew had existed on three biscuits and a dry herring each.

A few days later the *Honourable Artillery Company* and the *True-to the-Core* were out together after the new French steamer *Amiral Aube*, on her maiden voyage from Hamburg to her home port of Havre, had grounded on the Sunk sand on Friday January 13th. She was towed off by the tugs *Warrior*, *Burmah*, *Cambria*, *Columbia*, *Spray* and *Merrimac*, assisted by three lifeboats, after about 350 tons of coke and 60 tons of coal had been thrown overboard to lighten her—a task that occupied the Walton lifeboatmen for three whole days. The crew of the National boat was awarded £390 and the volunteer crew £325 for their share of the salvage work; in addition the *True-to-the-Core* was awarded a boat's share of £80.

This was only one occasion on which the lifeboatmen were away for days on end. In November 1900, the crew of the *True-to-the-Core* were absent from home for a week when they went out to help get the s.s. *Sardonyx* off the Kentish Knock. Six of the Walton men were taken off this steamship by the Margate lifeboat *Civil Service No. 1* when it was necessary to leave the vessel, because the weather blew up rough, and the *True-to-the-Core* ran down to Ramsgate for shelter. Very contradictory reports were current at the time, but the volunteer lifeboatmen seem to have done some worthwhile work, for they were awarded £240 for their services in the Admiralty Division the next year.

After 16 years, in which she saved 132 lives and helped to save six vessels from destruction, the *Honourable Artillery Company* was replaced by a new lifeboat. The crew chose a bigger and heavier boat of the Norfolk and Suffolk type, not unlike the *True-to-the-Core*, in place of the self-righting boat in which they had done such good work. The new boat, the *James Stevens No. 14*, was 43 ft. long and 12½ ft. in beam, had four water ballast tanks and was fitted with two centre plates.

The *James Stevens* lay afloat near the pier, like the *True-to-the-Core*, except when brought ashore for a refit in the summer, and the crew were ferried out from the pier by an old lifeboat, the former *Ramon Cabrera*, which had been stationed at Rhoscolyn (Anglesey) from 1878 to 1899.

One of her earliest services was to the Norwegian barque *Halden*, which grounded on the Gunfleet about four miles south-west of the lighthouse in a thick haze on November 29th 1901. She remained unseen and her signals went unnoticed until next day, when the Gunfleet lightkeepers spotted her and called out the Walton lifeboats. The *James Stevens* took off twelve of the crew and the *True-to-the-Core* later landed the other two, who remained behind to arrange for some of the ship's gear to be salvaged by Brightlingsea smacks. The wreck later went to pieces.

When the alarm was raised one January night in 1902 helpers went to haul the boarding boat alongside the pier. Up came the end of the 4-inch manilla haul-in rope, apparently cut. One man immediately threw off his oilies and boots and prepared to plunge in, despite the rough seas, the bitter cold and the darkness. He was stopped by the Coastguard chief officer, who refused to let him take the risk of swimming out alone, so he launched the pier boat from its davits and rowed to the ferry boat. Meanwhile, almost unnoticed by the crowd, three other men had gone down the steps into the sea to swim to the boarding boat 50 yards away. The four men were later presented with aneroid barometers in appreciation of their action.

The boarding boat was badly damaged when blown from her moorings and jammed against the pier in January 1903, and as a result the Clacton boarding boat was borrowed for some months until the former North Berwick lifeboat *Fergus Ferguson*, a 34 ft. self-righter, was sent as a replacement.

Sometimes when the weather was light the volunteer lifeboat crew would send their boarding boat, the *Peter Bruff*, ahead. With her lighter draught she could be pulled straight over the sands where the big lifeboat could not sail. The volunteer crew were successful in these tactics on September 27th 1904, when the *James Stevens* and the *True-to-the-Core* started nearly together, both spreading as much sail as possible to take full advantage of the light airs. The men in the *Peter Bruff* were employed by the captain of the grounded brigantine *Willie*, of Faversham, to take out a bower anchor and chain and to haul her off as the tide rose.

It is by no means easy to draw up an authentic list of services for a 'private' lifeboat such as the *True-to-the-Core*. The Institution had its own standards and the figures issued from its headquarters were accurate to a degree; the crew of a vessel which went on the sands on a foggy day in otherwise fine weather would be 'landed' by an Institution boat, but even the crew of a ketch barge run down by a steamer who rowed themselves to a nearby lightship and were picked up from there by a volunteer lifeboat went down in the records, such as they were, as 'saved'.

Some most exaggerated claims were made for the *True-to-the-Core*, and it is difficult to say just how many lives she really did save. Some of these exaggerated figures were published in the national Press with reports of a 32-hour trip made by the private lifeboat in December 1905. Both the *James Stevens* and her rival were rowed eight miles to the Sunk lightvessel on December 12th when the guns were heard through the fog, but although they searched for a long time nothing was found in need of assistance.

The *True-to-the-Core* had another lengthy trip just two years later. Friday, December 13th 1907, did its best to live up to what might be expected of it; the wind was 'through the Wallet' and the weather was very bad indeed, but the volunteer boat's crew mustered when it was heard that the Harwich lifeboat had gone out to a vessel burning flares. They sailed out to the Cork lightship and thence to the Shipwash. The wind was increasing all the time, and when the *True-to-the-Core* began turning to windward her crew really felt the force of it. The spray stung their faces like stones; they were drenched and frozen, but had a long way to go before they could find any kind of comfort. It took them four hours to close the land off Felixstowe. At daybreak the wind changed a point in their favour and they headed for the Naze.

Their thoughts must have been of home and warm beds—who can blame them—but when the spritsail barge *Havelock* was spotted at anchor in Dovercourt Bay, its sails in ribbons, the lifeboatmen changed course to investigate. 'Are you all right?' they asked. The skipper hesitated, as well he might, for he had been getting ready to hoist a flag in the rigging as a distress signal when he saw the lifeboat sailing towards him. The mate was at the pump, and down below was the skipper's wife and their five children. Eight of the lifeboatmen jumped aboard the barge as soon as the *True-to-the-Core* got alongside; the lifeboat's lugsail was passed to the barge and was rigged as a jury sail and they shaped course for Harwich harbour. Within five minutes the barge's foresail blew to pieces, and without it the barge dropped down to leeward much too fast.

The lifeboat sailed under the lee of the barge and, at the second attempt, a line was thrown to the barge. But as soon as the lifeboat took the strain on the towrope it carried away. Another attempt to take the barge in tow with a heavier hawser succeeded, however, and the lifeboat towed the barge into harbour.

This feat of towing a larger vessel under sail was not unique, but the sailing lifeboats had many limitations. If only the lifeboats had not been so dependent upon steam tugs the story of the *Indian Chief* might have been a happier one, and the coming of the internal combustion engine put a new tool into the hands of the lifeboatmen. Before a lifeboat could be fitted with a petrol motor, even experimentally, certain problems inseparable from the nature of lifeboat work had to be solved. The engine had to be protected from the effects of salt water, and the carburettor had to be so designed that it would still work with the boat standing almost on end or listing heavily.

By about 1905 the R.N.L.I. decided that engine development had reached a stage at which it would be worth experimenting with petrol motors for lifeboat purposes, and following encouraging experience with one of 10 h.p. fitted into an old self-righting boat, the *James Stevens No. 14* was chosen with two other boats of different design for further experiments. In 1905-6 she was fitted with a four-cylinder 40 h.p. Blake motor.

During the trials of the *James Stevens* water continually found its way into the carburettor and trouble was experienced with the clutch while, to quote the *Life-boat Journal*, 'The ordinary irritating little troubles well known to owners of motor-boats and motor-cars have been experienced in a more or less degree'.

In service, however, the *James Stevens* and her crew of Walton men soon proved the advantage of a motor. Even if the latter was not entirely perfect, and even if it was necessary to consider it only as an auxiliary, retaining sails and oars as the primary means of propulsion, it still enabled the lifeboat to accomplish tasks which would have been impossible a few years earlier. In August 1907, soon after her return to the station, the *James Stevens* towed the steam trawler *Westward Ho* off the Kentish Knock and the Faversham schooner *Mary Ann* off the

Gunfleet. She also assisted the barque *Adele*, also grounded on the Gunfleet, and at the same time gave help to a pleasure boat from Frinton whose motor had broken down.

Naturally the volunteer lifeboat found herself at a disadvantage without a motor, and as soon as subscriptions were sufficient the *True-to-the-Core* was sent to Lowestoft to be fitted with a 45 h.p. Brooke motor. The boat that had had a bigger sail than the old Institution lifeboat was now to be given a motor that was just that little bit more powerful than that of the new Institution boat.

While at Lowestoft the *True-to-the-Core* was in collision with a fishing vessel in the harbour and was badly damaged, but after repairs she returned to her station as a motor lifeboat, one of the few 'private' lifeboats to be so converted.

It was not long before the rivalry that had lasted nearly twenty years was renewed under the new conditions, and the *James Stevens* and the *True-to-the-Core* had a race under sail and power on June 5th 1910, when they chased a yacht which had picked up three men from a small boat out in Goldmer Gat. The *True-to-the-Core* won the race and returned the three men to their vessel, the Ramsgate smack *Valiant*, which had by then drifted ashore near the Naze. The crew had abandoned her when she ran on to the West Rocks, off Harwich, hoping to get aboard one of the other Ramsgate fishing boats which were in the area. The *Valiant* was eventually refloated and taken into Harwich by men from the Walton boat *Westward Ho*, a successful piece of salvage for the Walton men.

The superiority of the motor lifeboat was shown later the same year when the *James Stevens*, called to a vessel on the Barrow sand, rapidly overhauled the Frinton volunteer lifeboat and arrived at the wreck an hour and three-quarters before the sailing boat. On this occasion the *True-to-the-Core* had trouble with her engine, perhaps as a result of having got sand into the cooling system when she grounded on a call to the same vessel a few days before.

The final years of the Walton volunteer lifeboat are shrouded in mystery, for in 1911 there was a move for her to be taken over by the R.N.L.I. Tests were carried out by the Institution's officers at Harwich, where the boat went on the slip for survey, but the outcome seems to have been a refusal to accept her. It is said she was sold to Turkey to help start that country's lifeboat service.

Henry Britton, who had been apprenticed on a smack sailing out of the Colne as a boy and had later become part-owner of the Walton pleasure boat *Skylark*, a 32-tonner said to be capable of carrying 130 passengers, had a remarkable record of service. During his thirty years as coxswain he missed only one service in the lifeboat—and that was when he was recovering from injuries received in a collision between the *James Stevens* and the Harwich steam lifeboat. He retired in 1914 at the age of 66.

During the 1914-18 war the Essex lifeboats were kept busy assisting both allied and neutral vessels in trouble among the sands of the East Coast. One of the biggest wartime rescues was carried out by the Walton boat in 1917, when a morse message flashed from a destroyer to the Walton coastguard station called the lifeboatmen out on a hazardous and uncomfortable trip to the Longsand. Wind and tide were against them as the fourteen Walton lifeboatmen set off in the *James Stevens* on December 29th, but the motor took them safely round the northern end of the Gunfleet. Coxswain William Hammond was misled at first by a flare south-by-west, but he took the lifeboat down under the Longsand and eventually located the casualty, the General Steam Navigation Company's steamer *Peregrine*, which had run on to the north Longsand in almost the same spot that another vessel had broken her back only a couple of days before.

The sea broke steeply in the shallow water around the wreck, which was firmly wedged on the sands, and the lifeboatmen spent some time trying to get near her, but by 4.30 a.m. the lifeboat was within hailing distance of the stranded steamer. Captain W. Branthwaite hailed Coxswain Hammond from the bridge of the grounded vessel, telling him there were 59 passengers on board as well as the crew of 33.

'I'll rescue the women and children first and return for the crew!' bellowed the coxswain above the howl of the wind. 'Signal to the nearest Government patrol boat to take the passengers.'

While a signal was being made to one of the nearby naval vessels Coxswain Hammond brought the lifeboat alongside the steamer and took off the passengers, who included several children. He transferred them, 47 Frenchmen, 11 Belgians and an Italian, to a naval vessel which took them into Sheerness. Returning to the *Peregrine*, the lifeboatmen found she had swung round and her bows had become locked with the wreck of the Swedish steamer *Iris*, from which the Clacton lifeboat had carried out a difficult rescue not many hours earlier. This meant that the crew of the *Peregrine* had to clamber from their own vessel to the *Iris*, and thence to the lifeboat, not a speedy operation in such conditions.

Coxswain Hammond battled to keep the lifeboat alongside while the crew were taken off, but this proved most difficult in the sea that was running. While the boat was lying by the wreck a sea caught the rudder and the tiller swung over, flinging the coxswain against the mizzen rigging. The rigging kept him from going overboard, but Coxswain Hammond was badly bruised and shaken.

In spite of all this, the survivors reached the lifeboat in safety, bringing with them two cats. Nearly twelve hours after leaving her station the lifeboat turned for home, and three hours later the rescued men were landed in the backwaters. The lifeboatmen were nearing exhaustion, for they had struggled against mighty seas without food for fifteen hours.

William Hammond received the Silver Medal and Second Coxswain J. C. Byford the Bronze Medal for that service.

During 28 years of service the *James Stevens No. 14* was 'launched' 126 times and saved 227 lives. In 1928 she was replaced by a new boat, the *E.M.E.D.*, built at Cowes at a cost of £8,700—more than six times as much as her predecessor. This was one of the three boats of the Ramsgate type, 48 ft. 6 in. in length, with great beam, moderate draught, a flat floor and low end-boxes, designed especially for estuary work, the others being stationed at Ramsgate and Southend. In the 25 years she was at Walton the *E.M.E.D.* was very well liked by her crew.

In the 1930's the lifeboatmen began to find themselves faced with a new task. More and more frequently they were called out to assist yachts and small boats in trouble during the summer months, though at that time the problem of holidaymakers and inexperienced yachtsmen had not assumed the proportions it has assumed since the second world war.

The year 1930 was a busy one for the Walton lifeboatmen. The service record contains seven entries for that year, three of them services to pleasure craft. Three of the others, in March, were related in that they were all concerned with the stranding of a steamer on the Longsand—on March 15th the *E.M.E.D.* gave help to two boats from the collier *Corchester*, which was standing by the wrecked vessel, the s.s. *Arnent*, of Sunderland, an almost new steamer of 1,656 tons owned by James Westoll Ltd. On the following day and again on the 18th she took the captain and officials out to the *Arnent*, whose crew had been picked up and taken into Harwich by the L.N.E.R. steamer *Train Ferry No. 2*.

Late on April 11th the *E.M.E.D.* landed four men, the crew of the ketch barge *Mazeppa*, from the Kentish Knock lightvessel, to which they had rowed after their vessel had foundered as a result of being run down by the Norwegian steamer *Steingrin*. The *Mazeppa*, which was owned in Hastings, was bound to Rye with a cargo of coal.

On October 25th that year the lifeboat was called to the assistance of two men sailing from Walton to Clacton in a 15-feet lugsail boat, landing them at Clacton pier, and a little less than a month later took seven men and two women from the 100-ton auxiliary schooner yacht *Auriga*, which had gone foul of the Shipwash in half a gale. The *E.M.E.D.* was out again within hours of landing the people from the *Auriga* to tow in the yacht *Haliday*, found abandoned on the Buxey.

One October day in 1939 the sailing barges *Esterel* and *Yampa* were making their way along the East Coast on passage from their home port of London to Norwich with holds full of maize intended for the merchants of that old East Anglian city. The crews of the two barges were keeping their eyes open for enemy aircraft, for war had broken out a few weeks before and the barges were sitting targets for prowling raiders.

But it was the sea, the sailor's oldest enemy, and not German bombs, that was to cut short their voyage to Norfolk. An easterly gale sprang up when they were almost abreast of Orfordness and the two barges turned back, their skippers hoping to run into Harwich for shelter.

Both struck the Cork sand. The *Esterel* succeeded in getting off, but her sprit carried away and she was driven out of control towards the Essex coast, heavy seas washing over her and smashing the hatches. It was nine in the morning, and the tide was at half flood, as she grounded east of Walton pier in about six feet of water 100 yards from the shore. The heavy seas broke over her and then roared on towards the shore, smashing the beach huts lining the promenade.

Within minutes the local lifesaving team had erected their apparatus and one well-directed rocket carried a line over the wrecked barge. It seemed that the barge's crew would be hauled ashore safely in the breeches buoy, but then the sea played one of its nastiest tricks. Staggering under the onslaught of the waves, the barge swung round and the line got under her bottom. There was no hope from that direction.

The *E.M.E.D.* set off from her moorings under the lee of the pier some 50 minutes after the barge had gone aground, and a quarter of an hour later Coxswain Tom Bloom anchored the lifeboat to windward of the *Esterel* and began to drop down stern first towards the barge's bows. When off her bows he shifted the cable from the forward post to the starboard after bollard and steamed in under the lee of the barge, bringing the lifeboat's anchor cable tight across the barge's bows and so drawing the lifeboat's stern against the side of the barge.

Instantly lines were thrown from the lifeboat to the barge and made fast fore and aft. By keeping the lifeboat still steaming ahead the coxswain held her long enough for the master and his wife, the mate and their dog to jump into the lifeboat.

Now came the most dangerous moment of all. Fifty yards away the surf was boiling on the beach, and the lifeboat bumped the bottom several times as she rose and fell, staggering under the weight of water pouring across her decks and out through the relieving valves. The slightest mistake could put her ashore, and Coxswain Bloom dared not take her any nearer the beach. He moved her carefully astern, hauling on the cable at the same time, and so brought her clear of the barge's black bows. Once she was clear he gave the order for the cable to be shifted back to the forward mooring post and then went ahead on his engines and picked up his anchor. It was a splendid piece of seamanship.

The seas were too high for the lifeboat to return to the pier, so she made for the Backwaters. An hour and 40 minutes after putting out she was on her way up the Backwaters towards the yacht club hard, but the lifeboatmen's work was not finished. The lifeboat was met by a motor launch which took off the rescued crew, and then she set off again to buck her way into the teeth of the gale out to the Cork sand.

Fifty minutes earlier the Coastguard had received a signal from the Cork lightship that a vessel was in distress, and this message was passed on to Coxswain Bloom when the launch met the *E.M.E.D.* in the Backwaters. On the way out Coxswain Bloom could see a barge ashore on the Cork sand and he made straight for her. It was the *Yampa*.

Very heavy seas were running on the sands and the barge was completely submerged when the lifeboat reached her. Nothing could be seen of the crew, but the lifeboat circled the barge twice to make certain there were no men clinging to the rigging.

In spite of the way the lifeboat was being tossed about the lifeboat signalman, 'Toddy' Williams, managed to keep in touch with Walton Coastguards by means of the boat's searchlight, and the lifeboatmen also spoke to the Cork lightship and to a minesweeper, whose crew told them that nothing had been seen of the men from the *Yampa*. They searched the Wallet, hoping to come across the barge's missing crew, but nothing could they find. Walton Coastguard could give them no information, and at last they reluctantly turned for home, their work done.

Tom Bloom received the Silver Medal for that day's services, and second coxswain Jonas Oxley and signalman Williams were awarded the Bronze Medal, while the other five members of the crew got the Institution's thanks on vellum. One of those five was the assistant motor mechanic, Fraser Bacon, whose skilful management of the engines played a great part in the rescue of the *Esterel*'s crew.

After the wreck the *Esterel*'s cargo of maize was strewn up and down the beach, and on the following Sunday swarms of evacuee children scrambled over the barge's decks as scavengers gathered up buckets and sacks of grain. The *Esterel*, which like the *Yampa* had been built at Southampton in 1899, was later salved. She was again in trouble off the Essex coast a few years later, when her crew was taken off by the Clacton lifeboat, and after lying on the bottom for five months was raised and taken into the Colne, eventually returning to service as a motor barge.

By this time the Walton crew had already had a foretaste of the kind of call that was to bring them running many more times before victory in Europe was celebrated five years later. Even before the outbreak of hostilities, on July 17th 1939, the *E.M.E.D.* was called out when a Hawker Hurricane single-seater fighter from R.A.F. Debden crashed about four miles off Walton. It was not left to the lifeboatmen to pick up the pilot, who parachuted into the sea, for he was rescued by an R.A.F. seaplane from Felixstowe.

With the intention of taking part in the Dunkirk evacuation in May 1940, the Walton lifeboatmen took both their own boat and a new self-righting boat just being completed at Rowhedge Ironworks, but they were compelled to hand these over to naval crews. The whole story of the lifeboats' part in the evacuation is told in Charles Vince's *Storm on the Waters*, and it will suffice here to record that the Walton and Clacton boats worked in Dunkirk harbour, carrying men through the crowded darkness to the waiting ships outside. The Walton lifeboat was attacked by German aircraft on the way over, the officer in command of her was killed by a shell, and she eventually returned to Dover with a rope round her propeller. She went back to France after this had been cut away by a diver, and both the lifeboats were engaged throughout the evacuation.

In war-time it was impossible to use maroons to summon the crew and telephone messages or messengers on foot had to be employed to call the men. So it was that when, in the early hours of January 19th 1941, a policeman on duty on the front at Walton saw a flare just south of the town he had to call the second coxswain and bowman. The flare was also seen by one of the local lifeboat committee who lived on the front, and he telephoned the coxswain, who had the rest of the crew called. By the time the Coastguard had reached the secretary on the telephone the crew had assembled and were on their way to the lifeboat. The vessel in trouble was the Rochester barge *Martha*, a 92-ton spritty built at Rochester in 1897 and owned by F. T. Everard and Sons Ltd. She had lost her large anchor and had driven ashore on the Albion beach at Walton. The weather was appalling, a gale of wind, with snow, sleet and rain squalls, as the barge blew into shallow water, and by the time the lifeboat reached her the laden barge was being completely covered by the heavy seas. The Coastguard tried to use their rocket apparatus, but the rocket was blown back by the wind. The skipper had managed to let go his small anchor, however, and this held the stricken barge head to seas long enough for Coxswain Bloom to veer the *E.M.E.D.* alongside and take off the crew of three. The lifeboat hit the bottom several times, but the coxswain had her under control the whole time and in spite of the dreadful conditions he managed to complete the rescue without the boat being damaged. Coxswain Bloom and mechanic T. B. Brookes were awarded the Bronze Medal.

When it was reported on January 27th 1941, just a week after the rescue from the *Martha*, that a Swedish ship was drifting down Swin on fire it was the first the Walton men knew of a task which was to occupy them on and off for some weeks. The lifeboat was not required when the first reports came in about 1 a.m.—there was an Admiralty salvage tug in attendance—nor was the lifeboat called when the vessel, the steamer *Belgia*, of Gothenburg, drifted on to Frinton beach about 2.30 a.m. But soon after daybreak the Admiralty salvage officer asked for the lifeboat to take out fire pumps and firemen. Pumps were shipped in the boarding boat and the *E.M.E.D.* towed her off to the vessel, remaining there throughout the day. The following day the weather was too bad for the lifeboat to get alongside without great risk of damage, but she put out again on January 29th with firemen and a pump to finish the job of extinguishing the fire. The *E.M.E.D.* was out again on February 1st when tugs tried to refloat the *Belgia*, and she brought the salvage men ashore when the refloating attempt was abandoned after the tugs had dragged the ship about 50 yards.

By this time the tides had taken off, and it was impossible to refloat the vessel until the next spring tides. On February 14th the *E.M.E.D.* stood by while tugs pulled the *Belgia* off the beach, and then went alongside to take off the salvage men as the tugs set out to tow the ship into Harwich. 'This clears up this job,' the Walton honorary secretary noted in the record book as the lifeboat landed the salvage men on Walton pier.

The *E.M.E.D.* performed an unusual service four days later, when the Senior Naval Officer at Brightlingsea requested her services to take out naval ratings with a machine-gun to sink a floating mine about a mile south-west of Walton.

Five times in 1940 and seven times in 1941 the lifeboat was called to crashed aircraft as British and German airmen fought the air war in the approaches to London. Usually the *E.M.E.D.* searched and found nothing, but one day in May 1943, when there came a message that an aircraft had crashed off Frinton, the lifeboatmen found a German airman floating by his Mae West. He was dead.

Later in the war the emphasis changed, and the lifeboat crew was often called on to stand by when Allied bombers were returning from raids on the Continent. As the bomber offensive against the industrial centres of Germany developed, they were alerted more and more frequently; in one month the Walton crew stood by 17 times. They would sit around talking or playing cards in the lifeboat store, waiting either for the stand-down or for a message that would send them running to the boarding boat.

Peace returned, but there was no end to the lifeboatmen's work. On December 21st 1945, the Coastguard reported to the coxswain that an SOS was being made east-south-east of the Coastguard station, and Coxswain Bloom called the crew and the lifeboat was got away at once. It was already dark when the lifeboat set out, but the lifeboatmen managed to find the source of the SOS. Two naval motor fishing vessels, *M.F.V. 96* and *M.F.V. 611*, were ashore on the Gunfleet.

It was near the last of the ebb, and there was only just enough water for the lifeboat to get alongside one of the boats. The senior rating refused to leave his vessel, however, and asked the lifeboatmen to stand by. As the lifeboat was bumping heavily Coxswain Bloom decided to back off and stand by in deep water. It was not long before the two vessels again made SOS messages.

There was by now very little water around them, so the coxswain decided to take his boat round the end of them. In order to attempt to get the crew off the first he had to drive his boat straight at the vessel. As

Walton lifeboat carries out exercises with an R.A.F. Rescue helicopter—a helicopter crewman is lowered into the boarding boat City of Leicester. *(Photo: Frank Putman Ltd.)*

The spritsail barge Esterel *on Walton beach, 1939, after the crew had been taken out of her by the Walton lifeboat,* E.M.E.D. *She was salved and put back into service. (Photo: B. Norman)*

The radio ship Mi Amigo, *base for Radio Caroline, aground at Great Holland in January 1966. Those aboard were taken off by breeches buoy by the Walton L.S.A. company. (Photo: Eric Cheek, Essex County Standard)*

he got to her a heavy sea picked up the naval vessel and she came down right across the lifeboat's stern, smashing the steering gear. As the lifeboat backed away, with her rudder out of action, Coxswain Bloom found himself in a better position to get to the second vessel, so he went on and, steering by using his twin engines, got alongside her and took off the crew of six. Then he backed down to the first vessel and, still steering by manipulating the engine controls, manoeuvred alongside her and took off five men. It was no easy job getting clear of the wrecked m.f.v., but Coxswain Bloom contrived to get his lifeboat out into deep water. Then the crew had to fix up temporary rope steering, and when this was accomplished Coxswain Bloom decided to make for Brightlingsea, as he felt this was the only port he could hope to make safely with the damaged boat. After landing the rescued seamen the lifeboat was taken to Rowhedge for repairs.

Coxswain Bloom was awarded the Bronze Medal second service clasp for this rescue, during which the lifeboat was manoeuvred by her engines for four hours.

Little more than a month later the *E.M.E.D.* was taken out again to the help of a naval vessel. This time it was the minesweeper H.M.S. *Rhyl*, from which the lifeboat rescued 36 men.

The *E.M.E.D.* served at Walton until late in 1953, when she was replaced by a 46 ft. 9 in. Watson cabin type boat, the *Edian Courtauld*, which was provided by a gift from the late Augustine Courtauld, arctic explorer, yachtsman and member of the Institution's committee of management, in memory of his mother.

The old boat returned for a short period later as a reserve boat—while at Walton in 1955 she saved the steel sailing barge *Portlight*, one of Horlock's fleet from Mistley—but she eventually went far from the scene of her former exploits, being sold to Chile. Already more than 25 years old, she became the first unit of the Chilean lifeboat service, the Cuerpo de Voluntarios Botes Salvavidas.

Walton lifeboatmen braved nightmarish conditions in order to rescue the nine members of a Dutch coaster's crew on March 15th 1964. The lifeboat was called out shortly after 2.30 a.m. when news came that the 500-ton coaster *Merak* had run aground on the Shipwash sand while carrying a 658-ton cargo of steel from Ymuiden to London. Her skipper called for help in the belief that the vessel's rudder had been damaged, and that she had been holed.

The French tugs *Hardie* and *Jean Bart* raced to the scene and three British Railways cross-channel ships, the Harwich-Hook of Holland passenger steamer *Avalon* and the train ferries *Cambridge Ferry* and *Norfolk Ferry*, stood by. But these large vessels could not get near enough to assist the *Merak*, which was awash and buffeted by enormous waves as the *Edian Courtauld* was brought alongside by Coxswain Jonas Oxley. As the lifeboat rose and fell alongside the coaster one of the crew, 22-year-old Bobby Kemp, who was helping the Dutch crew into the lifeboat, was entangled in a rope ladder and slipped between the two vessels. It was a moment of near-disaster, but his colleagues pulled him back into the lifeboat uninjured.

The skipper and two members of the crew returned on board when the *Merak* refloated on the rising tide, and the coaster was towed to Gravesend by the Dutch tug *Poolzee*.

The lifeboat had only just returned to her station after this ten-hour trip to the Shipwash when the crew were told that a barge was in difficulties off Walton. This vessel got into Harwich under her own power. Then at midnight the lifeboatmen received a third call, when it was reported that the Trinity House pilot cutter *Penlee* was in trouble with flooding in the engine room. Again the vessel in difficulties reached Harwich in safety.

Jonas Oxley received the Bronze Medal second service clasp for this rescue. Shortly afterwards, on reaching retiring age, he handed over as coxswain to Frank Bloom, a son of Tom Bloom.

The coming of the 'pirate' radio ships to the Essex coast provided the Walton lifeboatmen with a new and totally unwanted occupation. Landing sick men and taking out doctors was one thing, but when the boat was called out to take supplies to a party of men who had been put on board one of the wartime forts on a sandbank off the coast the lifeboatmen took a poor view of the matter. They took an equally poor view of having to warn irresponsible youngsters that it was unsafe to attempt to get out to the radio ships on floating bedsteads, rafts and similar improvised contraptions.

A genuine lifesaving call to the m.v. *Mi Amigo*, which housed the *Radio Caroline* transmitting station, was answered on January 19th 1966, when the vessel broke from her moorings and blew ashore near Cheveau de frise Point, Great Holland, in a snowstorm. The crew had great difficulty in reaching their lifeboat, and a vellum of thanks was awarded by the Institution to Coxswain Frank Bloom and to each of his men for their courage, determination and skill in boarding the boat on this occasion. Although the disc jockeys of the *Mi Amigo* were brought ashore by the L.S.A. team, district inspector Lieut.-Cdr. D. B.

Cairns reported 'The coxswain and his crew showed courage, determination and skill in boarding the lifeboat in conditions of wind, sea and bitter cold which were the worst known for many years at this most exposed station'.

Coxswain Bloom won the Bronze Medal later in 1966 for a series of prolonged services extending over more than three days, the other members of the crew being given the thanks of the Institution on vellum. The maroons called the crew together at lunchtime on November 17th after the coastguard had reported that the Panamanian steamer *Ypapanti*, a vessel of 2,831 registered tons, was aground on the Longsand Head in a rough sea. Within a quarter of an hour of secretary Bob Oxley getting the message that the ship was aground the *Edian Courtauld* had slipped her moorings and was on her way to the casualty.

As she made for the Sunk lightvessel the lifeboat's radio brought news that the pilot cutter *Bembridge* and the Trinity House vessel *Vestal*, together with the Harwich train ferry *Cambridge Ferry* and another vessel, were standing by the *Ypapanti*, which was pounding as the seas struck her stern.

When the lifeboat arrived on the scene the heavy ground swell which was running in from the north-east was causing rough and confused conditions on the Longsand, and seas were breaking over the stranded steamer's starboard side and stern. The *Vestal* had already sent in one of her boats, but this was recalled when the lifeboat went in.

Coxswain Bloom brought the lifeboat along the Panamanian steamer's port side and secured her amidships, using the engines continually to ease the strain on the nylon securing rope as the heavy swell caused her to range considerably. At times she was on a level with the steamer's bulwarks, at others dropping down until it seemed that she must be smashed on to the hard sand below. The Greek master declined to abandon ship, so Coxswain Bloom decided to stand off as the weather was becoming worse.

As the lifeboat approached the edge of the bank a huge sea broke over her starboard side, bursting open the canvas wheelhouse screen and filling the wheelhouse. The after cabin was also awash, despite the fact that both doors were secured and the door athwartships closed.

The steam tug *Ocean Cock* arrived from Felixstowe, but the tide was falling and there was nothing she could do until the next high water, so she returned to Harwich harbour and Coxswain Bloom decided that as there was no immediate danger the lifeboat should return to its station. The German tug *Hermes*, which had arrived from its base at Yarmouth, stood by while the lifeboat turned for Walton, where the crew had a hot meal and changed into dry clothing before taking the lifeboat back to stand by the *Ypapanti* on the next flood. Still there was no sign that the crew wished to be taken off, so the *Edian Courtauld* returned to Walton once again.

At 8.42 a.m., a couple of hours after she had been moored under the pierhead, the coastguard told secretary Bob Oxley that the *Ypapanti* would not come off the shoal and that the master had asked for 13 of his crew to be taken off. Out went the lifeboat again.

On their third approach to the casualty the lifeboatmen found conditions had worsened since the previous day; the wind had veered to the north-east and the position of the ship was such that she now provided no lee. The main weight of the swell was breaking to the north-east of the *Ypapanti* and there was a secondary area of broken water off her bows. It seemed that the steamer had blown further up on the sand during the night.

Coxswain Bloom was anxious to take the lifeboat alongside as soon as he could as the very heavy breaking swell would be expending its main weight nearer to the steamer as the tide flowed. He asked the Trinity House vessel *Patricia*, which was nearby, to indicate the best way of approach—the officers on her bridge could see further than he could because of the extra height of the vessel's bridge. The coxswain, told that there appeared to be no gap and that he could expect a depth of five feet on the edge of the sand, decided to approach from the north without using the drogue to steady the lifeboat because of the risk of touching the ground and the drogue rope fouling the propellers.

She cleared the edge of the shoal safely but when abeam of the *Ypapanti's* bridge touched bottom in the trough of a wave. The following sea lifted her clear and on she went but as she turned to round the steamer's bow she again struck the ground heavily and lost way. Beam on to the sea, a wave broke over the whole length of her. The wave gave the hull some lift, and with the starboard engine full ahead and the helm hard to port the lifeboat slowly came round head to sea, two more waves breaking over her as she did so.

Coxswain Bloom brought the boat alongside the ladder just abaft the bridge and made her fast. She was rising and falling 15 feet as the waves ran along the steamer's side, and the engine controls had to be manipulated constantly to ease the strain on the mooring rope. In spite of this, eleven men were taken on board the lifeboat one at a time. Two more men intended to come off but refused at the last moment, and the captain refused to allow the remainder of the crew to be taken off, although conditions were getting worse.

The lifeboat passed round the stern of the *Ypapanti* into deeper water at 12.40 p.m., just 55 minutes after she had begun her run in to the casualty. The survivors were landed at Walton and the lifeboat was at once refuelled and made ready for service.

She was not needed again until the following day, November 19th, when at lunchtime Bob Oxley received a request for the remainder of the steamer's crew to be taken off. The lifeboat left her moorings at 1.15 p.m. and was once again near the casualty by three o'clock.

An hour and a half remained before high water and conditions around the steamer were extremely bad. The wind had increased to force eight, and seas were breaking over the *Ypapanti* and running along the decks. The lifeboat crew asked for a report on conditions on the steamer's port side, which could not be seen from the lifeboat. The master replied by radio telephone 'Not very hopeful'.

The *Ypapanti*'s stern was directly into the sea and the steamer provided no lee at all. Coxswain Bloom realised that if the lifeboat tried to take the remaining men off they would be exposed to the risk of serious injury. In the old days the risk would have had to be accepted but Coxswain Bloom was able to call on the services of an R.A.F. rescue helicopter.

When this left its base at Manston, Kent, at 3.40 there was still nearly an hour of daylight left, but darkness fell before the rescue was accomplished, so the lifeboat steamed slowly into the wind, her deck-light and searchlight illuminating the *Ypapanti*'s deck to help the helicopter crew.

The helicopter took two men off the steamer at the first lift and winched them down on to the lifeboat, then picked up three more men and transferred them with some difficulty to the lifeboat. The operation was completed by 4.55, and the lifeboat then returned to Walton.

For some time hopes were entertained of salvaging the *Ypapanti*. But the lifeboatmen reckoned she would never come off, and they were proved right, for she broke her back and eventually split into two sections, from which the sea washed the cargo of coke that she should have unloaded at Lisbon.

The lifeboat spent 24½ hours at sea in all—for the first three launches the crew were on duty for 26½ hours continuously—and the service ended with a classic example of co-operation between lifeboat and helicopter. The lessons learned in occasional exercises with the air-sea rescue helicopters of the R.A.F. had obviously been thoroughly learned.

The Walton lifeboat has proved in recent years that in spite of the advent of such modern aids to lifesaving as the helicopter and the introduction of fast boats such as that stationed at Harwich there is still a useful role to be played by the traditional lifeboats.

BOAT RECORD, WALTON

Years on Station	Length, Breadth, Oars/crew	Type, Weight, Cost	Year Built, (Official No.), Builder	Boat's Name, Donor, Authority
1884-1900	37′ 9′ 12 db	SR £394 10s.	1884 (31) Forrestt, Limehouse	*Honourable Artillery Company* Dramatic club of H.A.C. R.N.L.I.
1900-1928	43′ 12′ 6″ 12/	N & S 10t 16 £1,420	1900 (432) Thames Iron Works, Blackwall	*James Stevens, No. 14*[1] Leg. James Stevens, Birmingham R.N.L.I.
1928-1953	48′ 6″ 13′ 6/9	Ramsgate motor 21t 3 (22t 16 after 1932) £8,700	1928 (705) J. S. White Cowes	*E.M.E.D.*[2] R.N.L.I.
1953-	46′ 9″ 12′ 9″ -/9	W motor 22t 9 £29,687	1953 (910) J. S. White, Cowes	*Edian Courtauld* Augustine Courtauld R.N.L.I.
Boarding Boats				
1900-1903	33′ 8′ 10/	ex-SR	1877 (263) Woolfe, Shadwell	*Ramon Cabrera*[3] R.N.L.I.
1903-1922	34′ 7′ 6″ 10/	ex-SR	1887 (121) Woolfe, Shadwell	*Fergus Ferguson*[4] R.N.L.I.
1922-1932	34′ 8′ 10/	ex-SR	1894 (374) Forrestt, Limehouse	*R.A.O.B.*[5] R.N.L.I.
1932-	25′ 8′ 3″		1932 (BB31)	*City of Leicester*[6] Leicester Mercury Fund R.N.L.I.
Volunteer lifeboats:				
1894-1897	44′	N & S	1854 Beeching, Gt. Yarmouth	*True-to-the-Core*[7] Local subscription Salvage company
1897-1911	44′ 13′	N & S £437	1897 John Houston, Rowhedge	*True-to-the-Core*[8] Local subscription Salvage company

1 Fitted with motor, 1906
2 To reserve fleet, sold Chile
3 From Rhoscolyn
4 From N. Berwick
5 From Dungeness
6 Was at one time fitted with motor, but is now pulling boat.
7 Ex-*Royal Albert* from California, Norfolk
8 Fitted with motor, 1909. Sold c. 1911

SERVICE RECORD, WALTON

Honourable Artillery Company Lifeboat

1884	Dec.	25	Ship *Deike Rickmers*, of Bremerhaven, saved	25
1886	Dec.	4	Ship *Constanze*, of Hamburg, saved	19
1887	Dec.	2-5	s.s. *Capri*, of Liverpool, assisted in attempt to save vessel and rescued	22
1889	Nov.	10	Galliot *Edward*, of Skonevig, saved vessel and	2
1891	June	27	Brigantine *St. George*, of Guernsey, remained by vessel	
1892	Mar.	17	s.s. *Brighton*, of London, rendered assistance	
	Dec.	10	s.s. *Dilsberg*, of Glasgow, landed dead body	
1893	Apr.	9	Brigantine *Agnes Cairns*, of Fowey, assisted to save vessel and	7
	Nov.	10	Barque *Martha Brockelman*, of Rostock, landed a dog	
	Nov.	11	Barque *St. Olaf*, assisted to save derelict vessel	
	Nov.	20	Barque *Elphinstone*, of Christiania, saved vessel and	14
	Dec.	18	Schooner *Branch*, of Dartmouth, attempted to save vessel and rescued	5
1896	Dec.	4	Brigantine *Lord Strangford*, of Guernsey, assisted to save vessel and	7
1897	Feb.	18	s.s. *Paris*, of Hull, stood by	
	Feb.	24	Brigantine *Green Olive*, of Littlehampton, saved	8
	Dec.	5	s.s. *Banes*, of Bangor, stood by	
1898	Feb.	7	Ketch *Autumn*, of Hull, assisted to save vessel and	4
	Oct.	29	Brigantine *Stephen and Sarah*, of Whitstable, stood by	
1899	Jan.	14	s.s. *Amiral Aube*, of Havre, rendered assistance	
	July	25	Schooner *Little Dorrit*, of Fowey, saved	6
1900	June	28	Barque *Argo*, of Tvedestrand, assisted to save vessel and	13

James Stevens No. 14 Lifeboat

1901	Nov.	30	Barque *Halden*, of Fredrikshald, landed 12	
1904	Feb.	26	Schooner *Leif*, of Rudkjobing, assisted to save vessel	
	Sept.	27	Schooner *Willie*, of Faversham, stood by	
1905	Feb.	15	s.s. *Dunnikier*, of Kirkaldy, stood by	

A Temporary Lifeboat

1907	Jan.	26	Schooner *Demaris*, of Goole, stood by	

James Stevens No. 14 Lifeboat

1907	Aug.	7	Steam trawler *Westward Ho*, of Hull, assisted to save vessel	
	Aug.	22	Barque *Adele*, of Brevig, stood by	
			Motor launch, of Frinton, assisted vessel	
	Aug.	26	Schooner *Mary Ann*, of Faversham, rendered assistance	
1908	Aug.	24	Barque *Barden*, of Mariehamn, stood by	
1909	Mar.	5	Steam yacht *Varuna*, of New York, stood by	
	Sept.	4	Brigantine *Alroy*, of Hull, assisted to save vessel and	8
1910	Jan.	14	H.M. Tug *Diligent*, of Sheerness, assisted to save vessel	
	Apr.	12	Steam yacht *Sea Belle*, stood by	
	Apr.	25	Barge *Mary Jane*, of Rochester, saved vessel and	3
	May	19	Yacht *Gwenyth*, stood by	
1911	Sept.	30	Barge *Peace*, of Colchester, rescued a dog and saved	3
	Dec.	20	Barge *Charles Isabella*, of London, saved	7
1912	Apr.	27	Barge *Dorothea*, of Harwich, saved vessel and	2
	May	12	Barge *Dorothy*, of Harwich, stood by	
1914	Jan.	18	Brigantine *Eleanor*, of Yarmouth, saved	7
	Nov.	21	s.s. *Lilian Drost*, of Copenhagen, assisted to save vessel and	18
	Nov.	26	s.s. *Dalegarth*, of Newcastle, assisted to save vessel and	21
	Dec.	5	H.M.S. *Willett*, saved vessel and	13
	Dec.	29	Barge *O.L.S.*, of London, saved vessel	
1915	Jan.	17	Minesweeper H.M.S. *Eaton*, stood by	
	Mar.	30	s.s. *Glassalt*, of Aberdeen, assisted to save vessel	
	July	28	Ketch *Alexander*, of Lowestoft, saved vessel	

1915	Dec.	6	Ketch *Confiance*, saved vessel and	7
1916	Mar.	28	Schooner *Adolf*, of London, saved	4
1917	Dec.	29	s.s. *Peregrine*, of London, saved	92
1918	May	11	H.M. seaplane, saved seaplane and	3
1919	Mar.	11	Lighter *Cromer*, of London, saved vessel	
	Sept.	5	Motor schooner *Fionia*, of Copenhagen, assisted to save vessel	
	Oct.	1	Barge *Lord Kitchener*, of London, assisted vessel	
	Dec.	23	s.s. *Stanley Hall*, of Leith, assisted to save vessel and	17
1920	Nov.	1	Ketch barge *Lothian*, of Rye, saved	4
1921	May	20	Steam trawler *William Johnson*, of Hull, stood by	
	June	18	Barge *Siola*, of London, saved vessel and	4
	June	21	Fishing smack *Celebrity*, of Harwich, saved vessel and	2
	July	4	s.s. *Hani IV*, of Cardiff, stood by	
	July	5	Barge *Toots*, of London, saved vessel and	2
	Oct.	21	Barge *Leslie*, of Brightlingsea, stood by	
1923	Apr.	25	Barge *John Vidler*, of Rye, stood by	
	Aug.	30	Barge *Oceanic*, of London, saved vessel and	5
	Dec.	23	Longsand lightvessel, rendered assistance	
1924	June	6	s.s. *Pass of Brander*, of London, rendered assistance	
	June	10	s.s. *Baron Lovat*, of Ardrossan, stood by	
	Dec.	27	s.s. *Atlantic*, of Kubrickenburg, saved	6
1925	Aug.	7	s.s. *St. Abbs Head*, of Leith, stood by	
1926	Dec.	22	Fishing smack *Charlotte Ellen*, of Colchester, saved vessel and	4
1927	Apr.	6	Army pontoon *A.17*, saved pontoon	

E.M.E.D. Lifeboat

1928	Dec.	22	s.s. *Duris*, of Newcastle, saved vessel and	14
1929	Apr.	1	Motor speedboat *Wompy*, saved boat and	2
	Oct.	10	Barque *Plus*, of Mariehamn, rendered assistance	
	Oct.	11	s.s. *Otso*, of Helsingfors, stood by	
	Oct.	20	Barge *G.H.C.*, of Rochester, landed 2	
1930	Mar.	15	Two boats of s.s. *Corchester*, of London, rendered assistance	
	Mar.	16	s.s. *Arnent*, of Sunderland, took captain and Lloyds agent off to vessel	
	Mar.	18	s.s. *Arnent*, of Sunderland, took captain, Lloyds agent and surveyor off to vessel	
	Apr.	11	Barge *Mazeppa*, of Harwich, landed 4	
	Oct.	25	Small boat *Tim*, of Clacton, saved	2
	Nov.	22	Aux. yacht *Auriga*, of London, saved	9
	Nov.	23	Yacht *Haliday*, of Yarmouth, saved vessel	
1931	May	26	Steam trawler *Renovo*, of Grimsby, rendered assistance	
	Aug.	24	Aux. yacht *Marine*, of Ipswich, saved boat and	4
	Oct.	24	Small boat *Mary*, of Clacton, saved boat and	2
1932	July	6	Norwegian yacht *Sofie*, stood by	
	Sep.	11	Yacht *Pauline May*, saved yacht and	3
1934	Apr.	24	Barge *Lady Maud*, of London, stood by	
	May	30	Yacht *Argo*, rendered assistance	
	Dec.	9	m.v. *Titia*, of Dordrecht, assisted to save vessel and	5
1935	Jan.	27	Barge *T.T.H.*, of London, assisted to save vessel and	3
	June	30	Aux. yacht *Patronita*, escorted yacht into harbour	
	Sep.	26	s.s. *Ragna Forthon*, of Helsingborg, stood by	
1936	July	5	Yacht *Una*, of Brightlingsea, landed 4	
1938	Feb.	12	Barge *Chieftain*, of London, saved	3
	Feb.	13	Barge *Chieftain*, of London, saved vessel	
	Apr.	19	Barge *Gannet*, of London, saved vessel and	2

J. B. Proudfoot Lifeboat

1938	May	7	Barge *Lady Mary*, of London, saved vessel and	2

E.M.E.D. Lifeboat

1938	Aug.	1	Motor yacht *Betsey*, of Maldon, saved vessel and	4
1939	Jan.	7	Barge *Louise*, of Rochester, assisted to save vessel and	2

John and Mary Meiklam of Gladswood Lifeboat

1939	May	28	Motor yacht *Our Laddie*, of London, saved vessel and	13
	June	14	m.v. *Kentish Hoy*, of Chatham, saved vessel and	3

E.M.E.D. Lifeboat

1939	Sep.	3	Barge *Eureka*, of Harwich, rendered assistance	
	Oct.	3	Motor yacht *Herga*, of Guernsey, saved vessel and	6
	Oct.	14	Barge *Esterel*, of London, saved	3
	Nov.	21	s.s. *Terukuni Maru*, of Tokyo, saved ship's boat	
1940	Jan.	29	s.s. *Highwave*, of London, rendered assistance	
	Feb.	4	m.v. *Flores*, of Amsterdam, saved	6
	May	30	Lifeboat taken to Dover and, manned by naval ratings, used by Admiralty in connection with evacuation of Allied troops from Dunkirk	
	Nov.	19	s.s. *Folda*, of Leith, escorted vessel to safety	
1941	Jan.	19	Barge *Martha*, of Rochester, saved	3
	Jan.	27	s.s. *Belgia*, of Gothenburg, rendered assistance	
	Jan.	29	s.s. *Belgia*, of Gothenburg, rendered assistance	
	Feb.	1	s.s. *Belgia*, of Gothenburg, rendered assistance	
	Feb.	14	s.s. *Belgia*, of Gothenburg, rendered assistance	
	Feb.	18	Sunk floating mine	
1943	May	30	German aircraft, picked up body	
	July	9	Barge *Maria*, of Rochester, saved barge and	2
1944	Mar.	24	Two U.S. aircraft, saved a dinghy	
1945	Aug.	6	Man adrift in dinghy, saved dinghy and	1
	Aug.	23	Steam boat *Kimball Harlow* of London, stood by	
	Dec.	21	*M.F.V. 96* and *M.F.V. 611*, saved	11
1946	Jan.	28	Minesweeper H.M.S. *Rhyl*, saved	36
	Feb.	1	s.s. *Fort Massac*, of London, rescued cat and saved	1
	Feb.	3	s.s. *Fort Massac*, of London, took out ship's officers	
	June	3	Motor cruiser *Hu II*, of Bradwell, salved motor cruiser	
	Aug.	28	Aux. yacht *Zair*, of Maldon, gave help	
	Dec.	23	Barge *Glencoe*, of Rochester, saved boat and	2

Milburn Lifeboat

1947	Mar.	17	Aux. schooner *Ferjo*, of Westervik, stood by	

E.M.E.D. Lifeboat

	Apr.	23	Barge *Will Everard*, of London, saved	4
	May	24	Fishing vessel *Oceanic II*, of Ostend, saved	2
	June	8	Sailing boat *Oystercatcher*, of Burnham, saved	1
	Nov.	10	m.v. *Sodality*, of London, stood by	
1948	July	4	m.v. *Lenrodian*, of Sheerness, saved vessel and	6
	Aug.	15	s.s. *Rojo*, of Oslo, gave help	
	Aug.	17	Three fishing boats, gave help	
	Aug.	23	Motor cruiser *Asangyo*, of Newcastle, gave help	
1949	Feb.	27	Motor boat *Alnora*, gave help	
	Apr.	7	Motor barge *Success*, of Blyth, saved	4
	June	22	s.s. *Corcrest*, of Sunderland, saved	36
	Aug.	30	Aux. yacht *Arawatta*, of Maldon, saved yacht and	3
	Dec.	17	Barge *Thelma*, of Rochester, escorted vessel	
1950	Apr.	8	Motor yacht *Diana III*, of Richmond, gave help	
	Apr.	9	Pontoon type landing craft, saved boat and	2
	Apr.	25	Glider, saved glider and	1
	May	28	Aux. yacht *Mary Ellen*, of Walton, saved yacht and	4
	July	23	Naval cutter of H.M.S. *Ganges*, gave help	
	Aug.	13	Yacht *Norman*, saved a dinghy	
	Sept.	2	Rowing dinghy of Walton, saved dinghy and	4

Mary Scott Lifeboat

1951	Mar.	17	s.s. *Sodality*, of London, landed a sick man	

E.M.E.D. Lifeboat

	May	10	Yacht *Fortoyn II*, of Rotterdam, saved yacht and	6
	July	22	Yacht, saved yacht and	4
	Aug.	8	Yacht *Duet*, of Burnham, stood by	
	Aug.	12	Yacht *Siglinde*, of Maldon, gave help	
	Sep.	13	Yacht *Lassiette*, of Burnham, gave help	
1952	Jan.	18	German barque *Pamir*, stood by	

Mary Scott Lifeboat

	Mar.	29	m.v. *Seniority*, of London, stood by	

E.M.E.D. Lifeboat

	June	2	Yacht *Nola*, of Harwich, gave help	
	June	2	Dinghy of Walton, saved boat and	2
	Aug.	20	m.v. *Michel Swenden*, of Rotterdam and tug *Adrian Letzer*, of Flushing, gave help	
	Sep.	3	Yacht *Firefly*, of Gillingham, gave help	
	Nov.	6	Fishing vessel *Binney Stewart*, of Aberdeen, saved vessel and	2
	Nov.	7	Motor barge *Sway*, of London, saved barge and	3
	Nov.	7	Motor cruisers *Lady Betty* and *Celaiter*, saved boats	
1953	Jan.	11	s.s. *Campo Grande*, of Cadiz, gave help	
	Feb.	18	Motor boat, saved boat and	2
	May	6	s.s. *Duke of York*, gave help	
	June	18	m.v. *Bonnie Lass*, of Shoreham, gave help	

Thomas Markby Lifeboat

	June	29	Yacht *Alanna*, of London, gave help	
	July	23	Two persons marooned on sandbank, saved	2

E.M.E.D. Lifeboat

	Sep.	2	Ketch *Totland*, saved yacht and	6

Edian Courtauld Lifeboat

	Nov.	4	s.s. *Parkwood*, of Middlesbrough, gave help	
1954	Feb.	3	s.s. *Hera*, of Mariehamn, stood by	
	Mar.	31	s.s. *Brookside*, of Sunderland, escorted vessel	
	Aug.	4	Yacht *Dusmarie*, of Colchester, gave help	
	Sep.	20	Yacht *Schilleen*, saved boat and	2
			and persons cut off by tide, rescued	4
	Nov.	2	s.s. *Thelmsleigh*, of Hull, gave help	

E.M.E.D. Lifeboat

1955	Feb.	21	Barge *Portlight*, of Harwich, saved vessel and	2

Edian Courtauld Lifeboat

	June	20	Motor yacht *Shahjehan*, gave help	
	Aug.	6	Yacht *Bagatelle*, of Ipswich, gave help	
	Dec.	5	Yacht *Alexena*, saved yacht and	3
1956	Aug.	25	Fishing vessel *Dawn*, of Harwich, saved boat and	2
	Oct.	29	Yacht *Storm*, saved boat and	2
	Nov.	23	Airman baled out of aircraft but found dead in dinghy, landed a body	
1957	Aug.	2	Capsized dinghy, gave help	
	Sept.	1	Yacht *John I*, saved boat	
	Sep.	2	Yacht *Wild Duck*, of Burnham, saved boat and	2
	Oct.	22	Yacht *West Wind*, saved boat and landed 1	
	Oct.	26	Yacht *Porpoise*, gave help	

1958	July	3	Auxiliary barge *Ethel*, saved boat and	2
	Aug.	13	Aux. cutter *Nicolinda*, saved boat and	3
	Sep.	22	Fishing vessel *Dawn*, of Harwich, saved boat and	3
Cunard reserve Lifeboat				
1959	Apr.	7	Aux. yacht *Seaknight*, gave help and landed 2	
	Apr.	11	Aux. cutter *Rose*, saved boat and landed 3	
	Apr.	17	Yacht *Rigel*, saved boat and	3
	July	12	m.v. *Scarcity*, escorted vessel	
Edian Courtauld Lifeboat				
	July	22	Yacht *Rivale*, saved boat and	2
	Aug.	26	Yacht *Jabberwock*, gave help	
	Aug.	29	m.v. *Britta*, of Helsingborg, gave help	
	Nov.	13	Motor barge *Pepita*, escorted vessel	
1960	Jan.	2	Fishing boat *Cliffwind*, saved boat and	2
	May	27	Fishing boat *Alpha*, of Harwich, gave help	
	June	2	m.v. *Arasto*, of Norway, stood by	
Cunard reserve Lifeboat				
	June	9	Yacht *Carmen*, of Maldon, saved boat and	2
	Aug.	15	Yacht *Julia*, of Walton, gave help	
Edian Courtauld Lifeboat				
1961	May	6	Yachts *Pegasus* and *St. Barbara*, saved boats and	12
	May	20	Yacht *Tessa*, saved boat and	2
			Yacht *Tringa*, gave help	
	July	29	Yacht *Panacea*, of Harwich, gave help	
	Aug.	5	Motor cruiser *Carousel*, of Leigh, gave help and landed 2	
	Sep.	3	Ex-fishing smack *Quix*, saved boat and	3
	Oct.	21	Ex-Trinity House launch *Upway*, saved boat	
	Nov.	4	Dinghy, saved boat	
1962	Jan.	27	s.s. *Entopan*, landed sick man	
	June	24	Rubber dinghy, saved dinghy and landed 1	
	Aug.	4	Yacht, escorted boat	
	Aug.	6	Dinghy, saved boat	
Elizabeth Elson reserve Lifeboat				
1963	May	19	Yacht *Lisette*, of Heybridge, saved	1
	Aug.	24	Yacht *Evenlode*, of Burnham, gave help	
	Aug.	25	Yacht *Estella*, saved boat and	4
	Sep.	1	Motor boat, saved boat and	2
	Nov.	5	Raft *Walken Dick*, gave help	
Edian Courtauld Lifeboat				
1964	Mar.	15	m.v. *Merak*, of Groningen, saved	9
Elizabeth Elson reserve Lifeboat				
	July	19	Yacht *Galaxy*, of Mersea, gave help	
	July	27	Radio ship *Radio Caroline*, landed sick man	
	Aug.	1	Two dinghies, gave help	
Edian Courtauld Lifeboat				
	Sep.	13	Motor cruiser *Lady Jean*, gave help	
	Dec.	6	Yacht, saved boat	
1965	Jan.	6	Fishing boat *Maid Marion*, saved boat and	5
	Feb.	24	Radio ship *Radio London*, took out doctor	
Elizabeth Elson Reserve Lifeboat				
1965	Apr.	23	Radio ship *Radio London*, landed sick man	

Edian Courtauld Lifeboat

1965	May	3	Motor cruiser *Nan*, of Walton, escorted boat	
	July	3	Yacht *Two Chances*, of Wivenhoe, saved boat and	3
	Aug.	1	Two people cut off by tide, saved	2
	Aug.	15	Yacht *Campaigner*, of Maldon, gave help	
	Aug.	20	Radio ship *Radio London*, landed a sick man	
	Sep.	9	Radio ship *Radio London*, landed a sick man	
	Oct.	28	Sunk Head Tower (radio station), landed a sick man	
1966	Jan.	1	Trawler *Sheaf Field*, of Newcastle, landed a sick man	

Charles Cooper Henderson reserve Lifeboat

	Feb.	13	Radio ship *Radio Caroline*, landed a sick man	

Edian Courtauld Lifeboat

	Mar.	4	Tanker *Blandford*, of London, landed a sick man	
	Apr.	3	Aux. yacht *Souris*, of Maldon, saved boat and	4
	Apr.	11	Yacht *Wild Goose*, saved boat and	2
	May	15	Capsized catamaran, gave help	
	May	19	Barge *Asphodel*, of London, gave help	
	June	5	Catamaran *Ramrod*, saved boat	
	June	19	Speedboat from *Cheetah II*, saved boat and	2
			Rubber dinghy, landed 4	
	July	7	Motor cruiser *Leomar*, gave help	
	July	13	Yacht *Mona*, saved boat and	2
	July	24	Barge *British King*, gave help	
			Motor boat, no service	
	July	27	Aux. yacht *Marie Galante*, of Ostend, gave help, rescued	2
	Aug.	5	Dinghy, landed 4	
	Aug.	9	Sailing dinghy *Tiptoes*, saved boat and	2
	Aug.	12	Dinghy *Lady B*, saved boat and	1
	Aug.	28	Yacht *China Bird*, gave help	
	Aug.	29	Motor boat *Nil Desperandum*, gave help	
	Sep.	2	Yacht *Iolanthe*, gave help and landed 1	
	Sep.	4	m.v. *Corsea*, stood by	
	Sep.	9	Ex-naval pinnace *Julia Fisher* of Walton, gave help	
	Oct.	6	Motor cruiser *Sea Wolf*, gave help	
	Nov.	9	Sunk Head tower, landed 2	
	Nov.	17	m.v. *Ypapanti*, of Panama City, stood by	
	Nov.	18	m.v. *Ypapanti*, of Panama City, rescued	11
	Nov.	19	m.v. *Ypapanti*, of Panama City, landed 5	

WALTON VOLUNTEER LIFEBOAT

True-to-the-Core Lifeboat

1894	Aug.	27	Barquentine *Britannia*, stood by	
1895	Jan.	23	Schooner *Forest Oak*, of Sunderland, saved	4
	Feb.	3	Brigantine *Opah*, of Belfast, attempted to salvage vessel	
	Oct.	31	Schooner *English Girl*, of Fleetwood, assisted to save vessel and	5
	Dec.	26	Barque *Atalanta*, of Hamburg, assisted to save vessel and	18
1896	Apr.	12	Barge *Prosperous*, of Colchester, gave help	
	Sep.	12	Barquentine *Sunderland Packet*, of London, landed 6	
	Dec.	5	Smack *Emma*, of Grimsby, gave help	
1897	Oct.	25	Barque *Triton*, of Rostock, assisted to save vessel and	12

True-to-the-Core (second) Lifeboat

1898	Mar.	26	Ship *Pampa*, of Hamburg, assisted to save vessel and	25
	Oct.	30	Schooner *Stephen and Sarah*, of Faversham, saved	7
1899	Jan.	6-8	Schooner *Clementine*, of Nantes, assisted to save vessel and	5
	Jan.	13	s.s. *Amiral Aube*, of Havre, assisted to save vessel and	43
	Sep.	26	Steam yacht *Dispatch*, of London, assisted to save yacht and	3

1900	July	5	Schooner *Saxon Maid*, of Fleetwood, assisted to save vessel and	
			Boat *Myosotis*, of Frinton, saved	5
	Nov.	15	s.s. *Sardonyx*, gave help	1
	Nov.	25	s.s. *Count d'Aspremont*, of Swansea, assisted to save vessel and	11
	Dec.	28	Barge *Hope*, of Rochester, landed 2	
1901	Nov.	30	Barque *Halden*, of Christiania, landed 2	
1902	Jan.	6	Brigantine *Pet*, of Wisbech, assisted to save vessel and	6
1903	July	7	Barge *Intrepid*, of Ipswich, assisted to save vessel and	2
	Oct.	3	Schooner *Welcome*, of Preston, assisted to save vessel and	4
1904	Jan.	10-11	s.s. *Norham*, of Sunderland, landed salvaged goods	
	Feb.	10	Barge *Margaret*, of Harwich, gave help	
	Sep.	27	Brigantine *Willie*, of Faversham, assisted to save vessel and	7
1905	Feb.	15	s.s. *Dunnikier*, of Kirkaldy, stood by	
	June	1	Barque *Anna*, of Genoa, stood by	
	Nov.	26	Schooner *Alexander Nicol*, of Aberdeen, assisted to save vessel and	7
	Dec.	21-23	Schooner *Nancy*, of Goole, assisted to save vessel and	4
1906	June	22	s.s. *F. T. Barry*, landed 12	
	Sep.	17	Ketch *Malvoisin*, of Faversham, stood by	
1907	May	16	Pilot cutter No. 6, of Harwich, stood by	
	Dec.	14	Barge *Havelock*, of London, saved vessel and	8
1910	June	5	Smack *Valiant*, of Ramsgate, assisted to save vessel	

THE FRINTON VOLUNTEER LIFEBOAT

WHEN THE WIND brought the surf crashing on to the beach, and on to the sandbanks a few miles out, and stinging showers of rain or sleet were flung against the window panes by each successive squall, sharp eyes turned seaward from many a cliff-top vantage point along the Essex coast. Fifty or sixty years ago Frinton had its own little community of boatmen who in summer took holidaymakers for sea trips and on fishing expeditions and who in winter kept a eye open for wrecks on the Gunfleet.

Well known among these were the Cooks, who were closely linked with the Frinton volunteer lifeboats, the first of which was brought to Frinton by David Cook, an old Lowestoft beachman who moved with his family to Frinton about 1900, there to carry on business as a pleasure boat and bathing machine proprietor. When they first moved to Frinton the Cooks brought with them the *Courage Sans Peur*, an old Suffolk beach yawl, a long sweet-lined salvage craft of the type used by the beach companies for many daring life-saving and salvage missions off the East Anglian coast, and two smaller Lowestoft beach boats, the *Nil Desperandum* and the *Nellie Swift*.

The younger members of the Cook family became members of the crew of the Walton volunteer lifeboat *True-to-the-Core* and went out in her on many of her missions. William David Cook, a son of old David Cook, was in her on July 5th 1900 when she went off to the three-masted schooner *Saxon Maid*, of Fleetwood, which had run on to the Gunfleet. Two of his brothers, David Edward and John Hezekiah, and another Frinton man, John Jenner, also members of the volunteer lifeboat crew, missed the launch and so put off from Frinton in their own boat, the *Myosotis*.

When within a mile of the grounded vessel they realised they would not fetch the ship against the tide, which was running very strong and making a nasty sea on the sand. So they put her about, but just as the boat was in stays the jib sheets fouled, a sea caught her on the port side and in a moment she had capsized.

David sank with her, but kicking strongly he surfaced and looked round for a piece of floating timber or anything to which he could cling. 'Poor old Jack's gone!' He looked round to see who was shouting, and saw Jenner in the water not far away.

Both men kept themselves afloat until they were picked up, Jenner by the *True-to-the-Core* and David by a boat from the *Saxon Maid*, but there was no sign of Jack Cook, who, weighed down by sea-boots and oilskins, and perhaps stunned by a blow on the head as the boat turned over, had drowned helplessly. After being given restoratives and dry clothes on board the schooner the two rescued men were rushed ashore by the steam lifeboat *City of Glasgow*, which had come out from Harwich, while their friends worked at jettisoning some of the *Saxon Maid*'s cargo of coal. The vessel was towed off by the tug *Spray* on the next tide.

While at Lowestoft the Cooks had operated the former Caister lifeboat *Godsend*, which had been built in 1875 and sold out of the service in 1894 after having apparently spent two or three years inactive on Caister beach as a pleasure boat. The R.N.L.I. sold her in May 1894, to Charles Peachman, of Lowestoft, for £4 'subject to the usual conditions'.

These conditions were that the buyer would 'at once remove and not replace all lifeboat fittings, including aircases, relieving plugs and trunks, leaving the bare hull with thwarts and rudder; the fittings removed shall be destroyed forthwith as fittings, to the satisfaction of the Institution's officers; the colour of the outside paint shall be changed', and so on. The purchaser signed an undertaking to do all this within three months.

A 32 ft. surf boat, the *Godsend* was rigged with a single big dipping lugsail, unlike the larger boats of the Norfolk and Suffolk type such as the Walton R.N.L.I. boat *James Stevens No. 14*, which had a standing lug mizzen as well. Peachman kept her for about nine months and then sold her to David Cook for £12.

Brought to Frinton in October 1901, she was soon in use for salvage work, for in the following January she beat the *True-to-the-Core* in a call to the Wisbech brigantine *Pet*. Perhaps the Lowestoft men had a reunion on the Gunfleet that January day, for in command of the *Pet* was another Lowestoft man, a Captain Hubbard, who had with him his wife. The Walton men and the crew of the Frinton boat worked together in jettisoning the *Pet*'s cargo, refloating the vessel during the evening and taking her into Harwich.

There was one curious point about the stranding of the *Pet*, a 146-ton trader built at Dundee in 1863. She went aground within half a mile of where she had struck on a previous occasion about eighteen months earlier.

Renamed *Sailors' Friend*, the old lifeboat was used by the Cooks for salvage and life-saving work during the winter and as a pleasure boat during the summer. There must have been something about a trip in a lifeboat that attracted the Edwardian holidaymaker—perhaps it gave them a feeling of security. She was launched 24 times to assist vessels in distress between 1901 and her replacement by a new *Sailors' Friend*, built at Harwich by J. and H. Cann in 1907, when she was sold away to Skegness.

There seems to be no record of how many lives she saved during those years, but she certainly carried out some quite good work. She was in action on July 7th 1903, when she and the *True-to-the-Core* went off to the Ipswich barge *Intrepid*, bound to Woodbridge with chalk, and both crews shifted the barge's cargo and pumped her out, getting her safely into Harwich.

But many of her calls proved to be abortive ones. On June 1st 1905, the *Sailor's Friend* and the *True-to-the-Core* went out, followed later by the *James Stevens No. 14* plus a tug and the steam lifeboat from Harwich, when the Italian barque *Anna* grounded on the tail of the Gunfleet while being towed through the Gatway into the Wallet. The skipper refused all offers of assistance and the barque, which was bound to the Colne with a cargo of Argentine wheat, was towed off by the tug on the next flood. Two days later she was called to the opposite end of the Gunfleet to help a brig which had gone aground, but before she or the Walton lifeboats could reach her the stranded vessel set sail and sailed away.

In December that year the Frinton men sent a small boat away to get employment when the coal-laden schooner *Nancy*, of Goole, went on the Gunfleet. The *Sailors' Friend* followed with a bigger contingent—but the skipper would not at first allow the cargo to be jettisoned, with the result that it was two days before she was refloated. Then some of the Frinton men, aided by volunteer lifeboatmen from Walton, helped the schooner's crew of four to take her into Harwich.

However busy she may have been in the salvage business, the *Sailors' Friend* was not often called on to land shipwrecked crews. But there was the day in August 1906, when one of her crew, staring out through the morning mist, saw a vessel on the Gunfleet and called others, who

The Walton lifeboat Edian Courtauld *alongside the s.s.* Ypapanti, *aground on the Longsand Head in November 1966.* (*Photo: D. Royston Booth*)

The Walton lifeboat Edian Courtauld, *a Watson cabin lifeboat built in 1953* (*Photo: R.N.L.I.*)

went out in the *Nil Desperandum*. Finding that the vessel, the barquentine *Lanoy*, of Dover, was firmly fixed, Ted Cook remained on board her and sent the others back to Frinton for the *Sailors' Friend* in order to take off the crew and their belongings—William Cook, another of the same family, took charge of her in Ted's absence. There were at that time five Cooks in the crew.

Coming across the Wallet with the nine men from the *Lanoy* the *Sailors' Friend* ran into trouble. The rigging of the old boat proved unequal to the strain placed on it and the mast broke off about three feet from the heel, but the crew rigged a temporary mast and sailed her back to Frinton without further mishap. The rescued crew rode with the lifeboatmen in a torchlight procession held the same evening to raise money for a new *Sailors' Friend.*

'Mr. Cook's lifeboat' put off a few weeks later to assist the barge *Mabel* which had sprung a leak off Holland. The Frinton men helped to man the pumps and thus get the barge to Harwich.

These Frinton boatmen were not the rough, hard-drinking fellows of one-time popular imagination. They were all teetotallers, so when the new *Sailors' Friend*, a 40 ft. pulling and sailing lifeboat of the Norfolk and Suffolk type said to have been designed by old David Cook, was launched at Harwich on August 5th 1907, she was christened with a bottle of sea water. The boat, which had cost £360, was described at the time as 'the property of Frinton', with ownership vested in trustees and she was managed by a committee representative of various interests in the town.

Like her predecessor, she was a busy boat, being used for pleasure trips out to the Gunfleet lighthouse in summer as well as for more arduous journeys to the same area in the winter months. In the first four years of her life she made 27 or more trips to vessels in distress.

One of the longest services was performed in December 1910, when her crew spent 106 hours helping to refloat the Wilson liner *Antigone* from the East Barrow sand. They and the crew of the Clacton lifeboat, which had stood by for nearly two days before their services were accepted, spent two or three days heaving sacks of barley and tons of locust beans and nuts overboard. The beans and nuts were strewn on the beaches for miles, to the delight of the children. Two tugs helped to get the ship off, but they had left the scene when, the day after her refloating, the *Antigone* went aground again owing to the steering gear breaking down.

Out came the Frinton and Clacton lifeboats again, to stand by for another 32 hours or so. When the Clacton lifeboat returned home after the *Antigone* had refloated for the second time and had continued her interrupted homeward passage from Smyrna the crew had with them a tabby cat, presented to Coxswain George Grigson by the steamer's crew.

The Essex coast received a severe buffeting in April 1911, cliffs being undermined at several places and sea defence works at Frinton being damaged by the high tides and rough seas. On April 5th the Harwich steam lifeboat was called out three times, and that night the lifeboats from Frinton and Clacton answered signals from the Swin Middle and Mouse lightships and the Maplin lighthouse. The two boats had a rough trip, one of the Clacton lifeboatmen having his 'tarpaulins' torn off him by the wind, and then the Russian schooner to which they had been called got off the Black Tail spit without assistance.

As the *Sailors' Friend* was putting off from Frinton beach the mizzen outrigger carried away, but she drove out to the Swin Middle lightship under her mainsail alone. Coxswain Ted Cook decided to approach the lightship and ask for information and sent Cecil Bambridge, a 20-year-old carpenter, up forward with a light to attract the attention of the light-keepers. As they ploughed towards the tossing red-painted vessel the coxswain gave the word to lower the mainsail so they should not rush by before they could get news of the wreck they had been called out to help.

The sail came down with a run. As it fell, the wind caught the canvas and some part of the sail or its gear caught Bambridge and knocked him overboard.

A shout went up, and frantically the crew tried to bring the boat round in order to pick up their companion, who would be kept afloat by his cork lifebelt. Without the assistance of the mizzen it proved almost impossible to get her round; oars were hastily shipped on one side to help her round, but the boat dropped down to leeward and nothing more could be seen of the missing man.

For three and a half agonising hours they searched. Then they realised with mounting despair and horror that there was no longer any hope, and they set course for Queenborough, on the Isle of Sheppey.

Bambridge's body eventually drifted ashore on the Isle of Grain, and a few weeks later he was carried by the Frinton lifeboatmen to his grave in Frinton churchyard. The coffin was followed by the crews of both the Walton lifeboats and the Clacton crew.

Not unnaturally the death of one member of the crew was followed by the circulation of a rumour that the crew didn't really know their job, that most of them were not seamen and that the boatmen among them were taking an undue risk in putting to sea with inexperienced volunteers. Bambridge himself was a carpenter, but he had been in the crew for four years when he was lost.

It was difficult to keep such a private lifeboat organisation as the Frinton one in good shape financially, and at the annual meeting in 1914 reference was made to 'a certain disinclination in the town to come forward and share in the responsibility of keeping up the boat'. In fact there were some in Frinton who suggested that it was no part of the civic fathers' job to run a lifeboat service.

How little interest there was at this period is shown by the fact that only eight people attended that annual meeting. They included Coxswain H. Austin and one member of the crew. One is reminded of Mr. Micawber's pronouncements when one reads that income for the previous year had totalled £62 14s. and expenditure had been £62 2s. 4d. —but there were said to be several outstanding accounts which would turn that slender balance into a deficit.

The coming of war later in 1914 gave such men as those who manned the *Sailors' Friend* other work to do, and for a time it was almost impossible to find crews for the lifeboat. Some men went to the war and never returned, and for those who did come back to Frinton after the Armistice there was no lifeboat, for in 1917 the *Sailors' Friend* sank at her moorings opposite the old Beach Hotel. When she was raised she was sold for conversion to a yacht, and that was the end of the Frinton volunteer lifeboat.

BOAT RECORD, FRINTON

Years on station	Length, Breadth	Type, cost	Year built, Builder	Boat's name, Authority
1901-1907	33′ 7″ 10′	N & S £12 (in 1895)	1875	*Sailors' Friend*[1] David Cook (later Frinton Volunteer LB Society)
1907-1917	40′	N & S £360	1907 J. & H. Cann, Harwich	*Sailors' Friend*[2] Frinton Volunteer LB Society

[1] Ex Caister R.N.L.I. lifeboat *Godsend*. Sold to Skegness, 1907.
[2] Sold for conversion to a yacht, 1917.

FRINTON VOLUNTEER LIFEBOAT

Sailors' Friend Lifeboat

1902	Jan.	11	Brigantine *Pet*, of Wisbech, assisted to save vessel
1903	July	7	Barge *Intrepid*, of Ipswich, assisted to save vessel
1905	June	1	Barque *Anna*, of Genoa, stood by
	Dec.	21-23	Schooner *Nancy*, of Goole, assisted to save vessel
1906	Aug.	20	Barquentine *Lanoy*, of Dover, landed 9
	Sept.	16	Barge *Mabel*, saved vessel

Sailors' Friend (second) Lifeboat

1910	Dec.		s.s. *Antigone*, of London, assisted to save vessel

CLACTON

ONLY SIX AND A HALF MILES separate the lifeboat stations at Walton and Clacton, but even today there is plenty of work for both boats. The Clacton lifeboat's area stretches away to the south past the Colne and Blackwater estuary to the Maplin sand off the mouth of the Crouch; the next station to the south is Southend, some 30 miles away in the mouth of the Thames.

Although Clacton is now a thriving town with a settled population of 25,000, swelling enormously during the holiday season, Kelly's Directory in the 1870's credited the place with only twelve residents; a decade later it had 1,300 inhabitants. It was hardly known as a seaside resort when the R.N.L.I. sent a lifeboat to the town in 1878 at the request of the local residents. The lifeboat, a 34-footer named *Albert Edward*, was presented to the Institution by the Freemasons of England, who decided to take the very practical step of endowing two lifeboat stations, one of which was Clacton, as a gesture of thanks for the safe return from India of their Grand Master, the then Prince of Wales, later Edward VII.

The somewhat ornate brick lifeboat house which was built to accommodate this first lifeboat still stands in Anglefield. When it was built, at a cost of £510, the Institution spent an extra £39 on putting a fence round it to keep cattle out; the thought of any cattle coming within a mile of it today seems ridiculous. It ceased to be used as a boathouse when it was decided to keep the lifeboat on the pier, but the masonic emblems and the monogram AE on the tower still bear witness to its origin, though it is now the home of an electrical wholesaler's business.

This lifeboat, the first of three to be presented to the Institution for Clacton by the Freemasons, was not Clacton's first, for there had been a lifeboat of some kind supplied by Peter Bruff, the engineer, who had an interest in the development of Clacton as a resort. It was manned by the Coastguard, but details of its services, if any, are lacking.

The *Albert Edward* carried out her first rescue on May 23rd 1878, several weeks before the official launching, when she took off the crew of six men and three boys from the brig *Garland*, of Shields, which the lifeboatmen found breaking up on the Gunfleet. She was stripped by the Colne smacks as she went to pieces.

An outstanding service one Sunday morning in October 1881, earned medals from the French government for the crew of the *Albert Edward.* Called early in the morning, the lifeboatmen found both wind and tide against them, but Coxswain Robert Legerton decided to take the boat into the Spitway in the hope of getting a tow from a passing steamer. In the event they did even better, for when the steamer *Consett* took the boat in tow after another vessel had ignored the lifeboatmen's signals the skipper sent hot food across to the lifeboat on a line, and the refreshment put fresh heart into the lifeboat crew, who had been struggling at the oars for hours.

They found the French steam lugger *Madeleine* on the Maplin. Legerton let go his anchor and veered down towards her, but as he did so heavy seas drove the fishing boat higher on the sands and out of reach of the lifeboat. At a second attempt a huge wave hurled the lifeboat almost on to the deck of the Frenchman, smashing the rudder and otherwise damaging the boat, but Legerton and his crew hauled 16 men in one by one. It was no easy task: one man was almost swept away, and he saved himself only by grasping at the beard of John Greer, one of the lifeboatmen.

Just a week before this rescue the *Albert Edward* had experienced a slightly less hazardous trip to the Maplin. On this occasion the lifeboat was away from her station for 44 hours, the crew helping to get the billy-boy schooner *Ocean*, of Goole, off the sand and up to Poplar. As soon as the billy-boy berthed in the Thames one of her crew was seen by a policeman making off with a lifeboatman's coat, being afterwards sentenced to a month's imprisonment.

'She'll have a rough night of it,' onlookers predicted as the *Albert Edward* went out to investigate flares seen south-by-west of the town on January 23rd 1884. And a rough trip, and a tragic one, it was.

Under close-reefed canvas the lifeboat crossed the Wallet and sailed over the Gunfleet. There were a number of steamers about, so Coxswain Legerton ordered his second coxswain, James Cross, to burn a blue light to warn shipping of their presence and, if possible, prompt a reply from the vessel for which they were searching.

'While the blue light was being held by Cross two or three very heavy seas struck the boat, following in quick succession, one of which

Launching the Albert Edward *from one of the slipways on Clacton pier about 1890. The building of these slipways, from which it was customary for the pulling and sailing lifeboats to be launched with all sail set, greatly speeded up the launching of the boats which had previously had to be launched off the beach with the aid of a team of farm-horses.*

The Chapman lighthouse, one of four screw pile lights placed on the sandbanks off the Essex coast in the mid-nineteenth century to guide shipping. All are now disused, the Chapman light having been dismantled in recent years.

Fast as well as seaworthy—the 44 ft. self-righting lifeboat Margaret Graham, *built by Brooke Marine at Lowestoft for the Harwich station to a design developed by the United States Coast Guard and adopted by the R.N.L.I. after extensive trials in the 1960s.*

went into the sails and filled the boat, and she had not time to free herself as the seas flowed on so quickly,' said the coxswain later when making his report to the Institution. 'I put the helm down, but she would not answer it, she heeled over, and then turned right over to port. I called out to everyone to hold on to the boat, and not let go; she came upon her starboard broadside, and there remained.'

The coxswain found himself under water, tangled in the boat's gear. He managed to extricate himself after a struggle and dragged himself to the port side, where he found one of the crew named Kill. The sheets were foul and he took out his knife to cut them to allow the boat to right herself.

The anchor had fallen overboard, and this brought the boat up. As she came head to wind she righted herself, but two of the crew, James Cross and Tom Cattermole, had been washed away. Cross had been out in the lifeboat no fewer than 33 times and had assisted in the saving of 116 people, while Cattermole was on his seventh trip. Both left widows and children.

The lifeboat lay to her anchor until daylight, and the lifeboatmen could then see nothing within three miles of them. The seas were heavy, and the cold wind chilled the soaked men to the marrow. There was nothing to be done except to set sail for Clacton.

When they arrived home after 13 hours at sea some of the crew had to be lifted out of the boat. 'Poor Cross and Tom Cattermole are drowned' was the message that spread like wildfire through the town, which was cast into mourning within a very few minutes of the boat's return.

It later transpired that the vessel showing the flares was the Rochester barge *Jessie*, which lay at anchor about a mile west of the Heaps buoy with her sails blown to ribbons. The Harwich lifeboat *Springwell* went out to her the day after the disaster and assisted her into Harwich.

Yet when flares were seen from a steamer on the Gunfleet only two days later the Clacton crew mustered again and took the lifeboat out to help get the steam collier *Hawthorn* off the sand. And so it has always been; disaster has never daunted the men of the lifeboats.

Coxswain Legerton was awarded the second service clasp to his Silver Medal and the R.N.L.I. voted £450 towards the relief fund set up to help the widows and children of the dead lifeboatmen, as well as voting additional rewards to the crew for that service. The crew agreed to give the £1 that each was entitled to for the service to the relief fund.

One result of the disaster was that the Institution sent a larger boat to the station to replace the 34-footer. A 37-footer was sent to Clacton for temporary duty while a 39 ft. boat was being built. The new craft, also given by the Freemasons and like her predecessor named *Albert Edward*, was the first built for the Institution with a drop keel; perhaps that was why the crew said 'she sails like a yacht', after she had been out on her first service in October 1885, helping to get the Faversham brigantine *Eliza Box* off the Maplin sand.

Launching a lifeboat in the old days could be as hard and bitter a struggle as the long haul to the wreck. When the 'guns' called the crew the whole town would sometimes come down to the shore to help drag the lifeboat out of its house. So it was one September day in 1884, when the town was aroused from sleep about six in the morning by the signal and a crowd of about a hundred dragged the lifeboat from the boathouse even before the horses, borrowed from a local farmer, were on hand for launching.

Even with such help the launch was no easy one. As it was dead low water, the nine horses had a long haul over the sands and it was an hour before the lifeboat was afloat. Then the water was too shallow for her to get alongside the 57-ton schooner *Mystery*, of Portsmouth (built at Galmpton in 1854), which had gone on the Gunfleet. A 'Gospel ship', she had been stationed in Calais harbour for three months with a cargo of Bibles and had gone on the Gunfleet while on her way from Calais to Portsmouth—as one lifeboatman aptly remarked, 'Her name's the *Mystery*, and how she got where she did is a mystery to me'.

The lifeboatmen took her into Harwich harbour, where they were given a good meal and a Bible each by the colporteur's wife. The three seamen aboard the *Mystery* do not seem to have rated very highly in the Clacton men's eyes, not surprisingly, for they said the woman was 'the best man of the lot'.

Launching was made easier when, in 1885, slipways built on each side of the pier at a cost of £690 came into use, but lifeboat work was still hard indeed until the coming of motors. Sometimes it was made a little easier by a steamer, which would tow the lifeboat on its way as the *Consett* had done in 1881. Even so, work on the stormy North Sea was for fit men alone, and one of the Clacton crew, Benjamin Addis, paid with his life when he ignored the fact that he had been ill and insisted on going out in the *Albert Edward* in January 1888. The lifeboat went to the Norwegian vessel *Bjarne*, which had grounded on the Gunfleet, but her assistance was refused and the vessel later refloated.

Exposure in the lifeboat contributed to Addis's death a day or two later, and the Institution voted £50 to his widow, while pointing out severely that a lifeboat was no place for sick men.

The first three months of 1891 were busy ones for the Essex lifeboats. March came in with the spring flowers brought on by the warm pleasant weather. But by the seventh it was showing its lion's face, and on March 10th the Essex coast was experiencing 'almost a blizzard' when a boat came ashore near Giles's Gap, about three miles from Clacton pier.

In the boat was one man, who told the coastguards and Coxswain Legerton that he was mate of the 97-ton schooner *J. W. Bebell*, of Beaumaris. Owen Owens, for that was his name, told them his story as they drove him back to Clacton in a cab. He told them how the schooner had struck the Gunfleet late the previous night, how the seas constantly broke over the vessel, and how about midnight the boat broke loose.

Hearing the cook, Thomas Jones, call out for help, he ran to his aid and found the boat lying across his legs. Just as he reached the boat a wave broke over the deck, washing the boat overboard with him clinging to it. He tried to hold on to the schooner, but before any of his shipmates could come to his assistance he had lost his grasp and been swept out of reach.

He had no oars. His clothes he had to use to plug a hole, and his sou'wester to bale out the boat. And so it was that he was blown ashore thoroughly exhausted but still able to tell his story.

The *Albert Edward* went down the slipway only 17 minutes after Coxswain Legerton had fired the rockets to call his crew; it was the fastest launch made by the boat up to then, and a very creditable one for those days. A steamer gave the lifeboat a tow for some distance, and as the Clacton men approached they could see the *Duke of Northumberland* with the pulling and sailing lifeboat from Harwich in tow, as well as the *Honourable Artillery Company* from Walton, all making for the same spot. The Walton lifeboat anchored under the lee of the schooner, but Coxswain Legerton brought the *Albert Edward* alongside the wreck and took off one man, Hugh Owens. He told the lifeboatmen he had tried to get the captain into the rigging with him, but a huge wave had swept him off the deck.

Jerry Roper, one of the Clacton crew, climbed up the starboard rigging of the grounded schooner and down the other side, in spite of the fact that most of the ratlines had carried away, to recover the body of the fourth member of the crew, Rice Parry, who had died before the lifeboats arrived.

On the way back from the *J. W. Bebell* the steam lifeboat stood by a waterlogged Scandinavian timber brig for two hours, helping to keep her clear of the Longsand. Hardly had the Walton boat returned home than she was called out again, to the same vessel that the Harwich lifeboats had been standing by. The Walton men had a terrible trip, for heavy seas half-filled the boat time after time, drenching the crew.

Yet that was not the end. Within twelve hours of her return from the Longsand the *Honourable Artillery Company* was called out once more when the crew of the Gunfleet lighthouse reported that they had seen a ketch strike the sand and overturn; the cargo seemed to go out through the side and the vessel broke up. It took some time for the lifeboat to get to the wreck, which had drifted into the Barrow Deep, and by the time she arrived there was no sign of the ketch's crew.

Robert Legerton resigned as coxswain of the Clacton lifeboat in May 1891, on becoming piermaster at Clacton. In 13 years as coxswain he had helped to save 215 lives and to salvage 13 ships. He was succeeded as coxswain by Walter Schofield—his brother Charles was also in the crew—who gained the Silver Medal of the R.N.L.I. for his work in November 1893, when the Clacton lifeboat was out on service four times in three days and rescued 28.

Half an hour after setting off to assist a Norwegian schooner in trouble on November 20th a sudden squall carried away the foremast and broke the yard, but the lifeboatmen rigged up a jury mast using two oars on the stump of the foremast and succeeded in taking the casualty into Sheerness. There Coxswain Schofield asked for a tug to take them to a barque they had seen on the Barrow sand, and the lifeboat was towed as far as the Swin Middle. They found the vessel on the Barrow was deserted, so set off back to Clacton, but on the way they saw another vessel near the Whitaker Spit with her gear hanging over the side and the rudder gone. Hailing her six times and getting no answer, they continued back to Clacton, but were just leaving the pier after mooring the lifeboat when distress signals were seen from this same vessel.

So off they went again, to rescue the captain and ten men from what proved to be the *Harald Haerfagre*, a barque which had left Northfleet several days before and had brought up in the Swin when forced back after losing her maintopmast, foretopgallantsail and mizzen topmast off the Leman and Ower sand, abreast the north Norfolk coast.

Next day the lifeboat made a 15-minute dash to the Danish barque *Anna* on the Gunfleet, standing by her until the evening, when the captain decided to abandon the ship as her back was broken and there was twelve feet of water in the hold. By the time the crew of ten had transferred to the lifeboat she must have resembled Noah's Ark, for they brought with them not only their sea chests but a large dog, three cats and a canary. The *Albert Edward* was being towed back to Clacton by the *Merrimac* when the crew saw distress signals from the Swin Middle. The lifeboatmen waited only to eat a few sandwiches and to drink some coffee before putting out again, but all they found was a steamer riding safely.

In 1901 the second *Albert Edward* was replaced by a new lifeboat of the same name, a 45 ft. boat built by the Thames Iron Works. This was kept on the pier during the winter ready for launching down a slipway, but each summer she was moved to permanent moorings off the pier-head in order to provide more room on the pier for holidaymakers, an old lifeboat being provided as a boarding boat. This latter, a Rubie-type self-righter, figured in the annual Lifeboat Saturday processions at Clacton, being drawn on a carriage through the town and often being launched afterwards to carry out a mock rescue.

One of the new lifeboat's early tasks must have been one of the dirtiest for her crew. Launched from the pierhead at night on February 2nd 1902, she helped to get the steam collier *Dungonnel* off the Black Tail sands. When they arrived back after jettisoning 80 tons of coal the Clacton lifeboatmen looked like a group of coalminers.

George Grigson, who became coxswain of the Clacton lifeboat in 1907 after seventeen years as a member of the crew, was one of the few natives of Clacton at that time, having been born in one of the two or three lonely wooden cottages which in the middle of the nineteenth century stood on the shore where Clacton-on-Sea was developed. Like his father, who hailed from Brightlingsea and was a member of the lifeboat crew from 1883 until three years before his death in 1902, he was a fisherman until in 1919 he left for Maldon to become a fish merchant.

Several members of the family served in the Clacton lifeboat besides George Grigson and his father. His brother Harold joined the crew in 1902, the same year as a cousin, Norman Wilson; another cousin, Ernest Wilson, was already in the crew.

Grigson was coxswain when the Clacton lifeboat was fitted with a motor. While the *Albert Edward* was away for this to be fitted the Clacton station was supplied with a reserve boat, the *Christopher North Graham*, but the crew decided she was unfit for service after one practice launch, and when signals were seen they refused to go out, leaving it to the Walton men to save the skipper and his wife and the mate from the Rochester barge *Peace*, which was dragging her anchors towards the Maplin. The Walton boat had to take the rescued crew to Sheerness, so that left only the Harwich and Frinton boats to put to sea when a call came the following night; they carried out a search but found nothing. The Clacton men subsequently chose a lifeboat of the Liverpool type, the *Charles Deere James*, to take the place of the boat they had been given.

Not only was the Clacton lifeboat one of the first R.N.L.I. craft to be fitted with a motor, she must also have been one of the first to launch to the assistance of an aircraft. She was called out on April 1st 1914, when a seaplane flying from the Isle of Grain to the naval air station at Felixstowe, on the Suffolk side of Harwich harbour, crashed off St. Osyth beach. The pilot and navigator were taken on board a passing barge and were later transferred to the *Albert Edward*, which towed in the wreck of the seaplane, using her sails instead of the motor in order to reduce the strain on the remains of the aircraft. Two other seaplanes landed near the lifeboat as she made her way slowly back to Clacton, where the wrecked plane was hauled ashore. Just over three weeks later another seaplane carrying Winston Churchill on a visit to the naval flying school at Harwich was forced down off Clacton. The *Albert Edward* did not have to rescue the First Lord of the Admiralty, but she was later called to assist a naval motor launch sent to recover his aircraft after he had left in a second seaplane sent from Harwich, and she had to tow both motor boat and seaplane to Felixstowe.

On occasion, lifeboatmen have shared the horrors of shipwreck with those they sought to save. Some of the Clacton crew were put in this position one September night in 1915 when the *Albert Edward* was called to the Guernsey barque *Leading Chief*, which had run on to the West Sunk. The crew of nine had begun jettisoning part of her cargo of ore so as to lighten their vessel, but had to take to the poop as the *Leading Chief* settled on the sand. The water crept above their chests as the barque sank, and just as the lifeboat neared her the mainmast went overboard. The tangle of rigging almost trapped the lifeboat as if in a net when the mast crashed down, but undeterred the lifeboatmen struggled to reach the nine men on the poop.

One man was washed out of the lifeboat, and it was only after a long and difficult struggle that his companions hauled him from the icy water. Then some of the lifeboatmen boarded the wreck in an attempt to get to the crew, and were marooned when the rising seas washed the lifeboat away from the *Leading Chief*. Second Coxswain Jesse Salmon took command, for Coxswain Grigson was one of the men left on the wreck.

Nine times he drove the lifeboat at the wreck and at last he succeeded in taking her right across the submerged deck so that the lifeboatmen and the barque's crew could clamber on board. As they drew away the *Leading Chief* was beginning to break up.

On the return journey the lifeboat's engine was out of action—perhaps the raffle of rigging had fouled the propeller. Under sail, beating into the teeth of the wind, the lifeboat made slow progress towards her station, which was not reached until the following morning. By then both rescued and rescuers were in poor shape and in great need of the hot food awaiting them.

Both coxswain and second coxswain were awarded Silver Medals for this rescue, while Grigson gained a third Silver Medal in 1918 for a rescue from a steamer on the Longsand, thus ranking with Coxswain Legerton, who received the third clasp to his Silver Medal on his retirement. Jesse Salmon received the Bronze Medal for the same service.

Mountainous seas were being whipped up by a biting easterly wind when a radio message from a naval vessel gave news of a Swedish steamer needing assistance on December 27th 1917. The message brought out Clacton lifeboatmen and the *Albert Edward* ploughed her way out to the Longsand, where a number of naval vessels were standing by the 1,254-ton steamer *Iris*, of Norrköping. Among these naval vessels were the paddle steamers *Clacton Belle* and *Walton Belle*, known in more peaceful times as popular steamers on the east coast and at this time employed on minesweeping duties.

Coxswain Grigson hailed one of them, learning that the *Iris* had gone aground at the north end of the Longsand, and that she was too firmly fixed for the minesweepers to get her off in the conditions prevailing. All day the *Albert Edward* stood by, and when in the afternoon the weather worsened Grigson offered to take off the crew, but his offer was refused. A little later, however, the lifeboat drew alongside the stricken ship so that the crew could lower their belongings into her.

The confused seas on the sands flung the lifeboat about and the spray swept over the crew, stinging their faces and running down inside their oilskins, freezing them to the marrow. It was too uncomfortable for them to remain near the ship, so they moved away and anchored in the vicinity.

The wind increased and the weather became steadily worse during the night, but still the lifeboat remained at anchor, watching over the doomed vessel. Suddenly a flare lit the inky blackness of the early morning; the Swedish captain, realising at last the hopelessness of his position, was asking to be taken off.

The lifeboat closed in, and Coxswain Grigson manoeuvred close to the steamer so that the 22 Swedes and the English pilot could jump one by one into the boat as she rose on the crest of each wave. The *Iris* was breaking her back on the hard, unyielding sand as the lifeboat turned for home, her crew hoisting to the masthead the red flag which told friends ashore that a rescued crew was aboard. Their difficulties did not end with the rescue, for when they arrived back at Clacton pier the heavy seas made it difficult to bring the lifeboat alongside, and she was damaged against the timbers of the pier. The crew had been afloat for 36 hours, soaked to the skin the whole time, with no proper food, and were all well-nigh exhausted.

The year 1925 was a busy one for the ageing *Albert Edward*; the service boards record that in the year she saved seven lives, but they do not reveal the whole story, for the lifeboat was called out several times on errands which did not result in any useful service—as is the case every year. On February 11th 1925, however, when called to the West Barrow she performed a most useful service, saving four men from the 57-ton ketch *Rosina*, of Plymouth, which had dragged her anchor and grounded on the sand while on a voyage with 120 tons of china clay from Par, in Cornwall, for London. The crew had burned everything they could lay hands on in an attempt to attract attention to their plight.

When the *Albert Edward* arrived her crew saw the *Rosina* up on the sand, with the Margate lifeboat aground about 200 yards to leeward. The *Albert Edward* got alongside and the ketch's crew were dragged on board, but there was nothing to be done to save the ketch.

A message from the Barrow Deep lightvessel took the *Albert Edward* out on July 29th to the north-eastern part of the Longsand, where the French topsail schooner *Walkyrie*, of Lannion, had gone aground. The seas were rough in the shallow area round the stranded vessel, which was on a voyage from Brittany for Aberdeen with 75 tons of onions, but Coxswain Charlie Ellis took the lifeboat close in and hailed her. Two French flags which had been run up as a distress signal were still flying, but he got no reply from what seemed to be a deserted ship, so he took the *Albert Edward* alongside and one of the crew went on board to investigate. On a bunk in the cabin he found a man, unconscious and with a bandage round his head; he was the only occupant of the schooner.

It was later discovered that when the *Walkyrie* went aground the other three members of the crew and the ten onion-sellers who were passengers had taken to the boat, being picked up by a steamer which landed them at Gravesend. The man in the bunk was the skipper, who had been injured by a swinging boom nearly three weeks before.

The lifeboat stood by for several hours until the schooner refloated on the rising tide, then Second Coxswain Frank Castle and four other lifeboatmen sailed her into Harwich. Once there the French skipper was able to receive treatment for his injuries.

The *Albert Edward* came to the end of her service at Clacton in 1929, when she was replaced by a new motor boat, also of the Watson type. This was the *Edward Z. Dresden*, a 45 ft. 6 in. boat built at East Cowes. The *Albert Edward* was transferred to Aranmore, where she served for a further three years before being sold out of the service. She eventually became a yacht, was taken over by the government at the beginning of the second World War, and did not disappear from Lloyds' Register of Yachts until about 1952, by which time she had been afloat for half a century.

The big sailing barges that once sailed to the Tyne after coal, over to the Continent and round to the westward, sometimes got into trouble among the sands of their native coast. The story of the Walton lifeboat's services to barges in 1939 and 1941, both of which earned medals for Coxswain Tom Bloom, have already been told, but there were many other occasions when the lifeboats went to the aid of coasting boomies and spritties.

Sometimes both the Walton and Clacton boats were needed to bring in a big barge. In February 1935, for instance, they both went to the aid of Samuel West's mulie *T.T.H.*, which had knocked off her rudder on the Gunfleet and had lost her boat in heavy seas. They brought the helpless barge into the Colne, the *E.M.E.D.* lashed alongside to port and the *Edward Z. Dresden* made fast on the starboard side.

In January 1939, the *Edward Z. Dresden* tried to get the London and Rochester Trading Company's spritty *Louise* to a safe anchorage when

The Frinton volunteer lifeboat Sailors' Friend, *built in 1907 to replace the former Caister lifeboat* Godsend *which had operated from Frinton for several years as a private lifeboat—a photograph taken at the naming ceremony on Frinton beach in August 1907.* (*Photo: V. G. Scammell*)

The Walton motor lifeboat E.M.E.D. *lashed alongside the mulie barge* T.T.H., *assisting her into the Colne in 1935 after the* T.T.H. *had knocked her rudder off on the Gunfleet sand. Also assisting, on the starboard side, is the Clacton lifeboat* Edward Z. Dresden. (*Photo: N. Harrison*)

The Clacton lifeboat Albert Edward *at her winter station on Clacton pier in 1900. She was launched down a slip, the top of which is just visible ahead of the boat. (Photo: East Essex Gazette)*

the barge lost both anchors and, with her steering gear smashed, became unmanageable off Holland Haven. She could not handle the heavy vessel on her own, however, and had to call for the help of the Walton boat; between them the two lifeboats towed the 49-year-old barge to Harwich.

A gale was blowing from the north-west, knocking up a very rough sea, and it was raining in squalls when at half past three on the morning of November 19th 1939, the coastguard reported seeing Very lights four miles south-west of their lookout. It was at that time of early morning when one's vitality is at its lowest, but within half an hour the crew had assembled and the *Edward Z. Dresden* had gone down the slipway.

As day broke the lifeboatmen found the Tollesbury stowboating smack *Charlotte Ellen*, which had been out fishing for sprats with a crew of three. She was anchored in a very dangerous position at Swire Hole with the sands close under her lee and was quite helpless. Her stowboat gear was standing out 15 ft. on the starboard side, so it was impossible for the lifeboat to approach on that side. Coxswain Ellis had to bring the lifeboat alongside on the port side, easing her between the smack and the sands, no easy task in the heavy seas, to get the three men off.

When the lifeboat arrived back at her station the coxswain learnt that the smack was not the vessel that had been firing the Very lights, so off went the lifeboat again. This time they found the barge *Lorna*, deeply laden with cotton seed, at anchor between the South Whitaker and Middle sands, the seas breaking right over her deck. The coxswain attempted to take the lifeboat alongside to take off the two-man crew, but the barge swung so violently in the raging seas that he had to back away. With the casualty swinging dangerously from side to side, Coxswain Ellis dared not approach again. It was useless to fire the line-throwing gun, for with the seas breaking right over the barge's deck her crew would never have been able to fix the rope and block so they could be hauled by breeches buoy to the lifeboat.

The coxswain took the only course left to him. He ran the lifeboat right on to the deck of the barge and lifeboatmen seized the skipper and mate in the brief interval before a heavy sea washed the lifeboat back off the deck. The lifeboat's stem had been damaged when it smashed into the barge's hatches, but its crew were saved.

For these two fine rescues Coxswain Ellis was awarded the Silver

Medal, his second coxswain and the motor mechanic both being awarded the Institution's thanks on vellum.

Clacton lifeboatmen had had experience of aircraft in distress during the first World War, but this was little preparation for the work that had to be performed during the 1939-45 war. On the second day of 1939 the lifeboat crew turned out because a flying boat was reported to be down in distress, but it was later learned that the aircraft was safely at anchor and the lifeboat was recalled by radio, and in July both Walton and Clacton lifeboats were called when an aircraft crashed between Walton pier and the Gunfleet lighthouse, but the pilot "baled out" and was picked up by an aircraft taking part in the search.

In the next six years the Clacton lifeboatmen were called out 29 times to search for survivors from crashed aircraft, but usually they found nothing; sometimes wreckage or a patch of oil was discovered but no sign of life.

On one occasion in 1941 both the Clacton and Walton lifeboats, plus the motor boarding boat then at Walton, put to sea when a British fighter came down in the sea, but the pilot was rescued by a barge. A few days later the reserve lifeboat *J. B. Proudfoot*, which was then on duty at Clacton, picked up a Polish pilot from the sea after his bomber had crashed in the early hours of the morning. The lifeboatmen found the bomber's dinghy, but a search revealed nothing of the rest of the aircraft's crew.

In September that year the *Edward Z. Dresden* helped to rescue three men from a British bomber which had crashed on the Gunfleet. When the lifeboat arrived the airmen were standing on the wings, but the water was too shallow for the lifeboat to approach, so the airmen paddled towards the lifeboat in their rubber dinghy. They were picked up, not by the lifeboat, but by an air-sea rescue launch from Brightlingsea, the lifeboat staying at the scene to help prepare for the recovery of the aircraft later.

The *Edward Z. Dresden* went to Dunkirk with other lifeboats at the end of May 1940, working with the Walton boat in Dunkirk harbour. While the evacuation was going on across the Channel, preparations for defending this country against invasion were being pressed forward with all speed, among the defence measures taken being the breaching of all piers to make them useless for the landing of troops. Thus when the *Edward Z. Dresden* returned from Dunkirk it was no longer possible for her to operate from Clacton pier and the station was closed for a couple of months. The lifeboat went first to Rowhedge for overhaul and was then stationed at Brightlingsea for three months until it was decided to bridge the gap in the pier and again operate from Clacton.

At this period the lifeboat was in the thick of the war at sea. She went out from Brightlingsea on November 7th 1940, just before her return to Clacton, after the naval drifter *Reed* had been mined off Holland Point. There was only one survivor, picked up by a speedboat, and he died in hospital after being landed by the lifeboat. Next day the Brightlingsea naval base reported that two steamers were sinking in the West Swin and the *Edward Z. Dresden* went out to help the s.s. *Fireglow*, a London collier, which had been hit forward by a bomb and raked by machine-gun fire which had smashed all the ship's boats. The other vessel had already sunk when the lifeboat reached the scene. The *Edward Z. Dresden* escorted the *Fireglow*, which had been taken in tow stern first by a tug, to Southend, where she landed naval officers from the tug and the master of the *Fireglow*, then put the master back on his vessel before returning to her station.

Ten days later the lifeboat rescued seven men after F. T. Everard's motor coaster *Ability* had struck a mine and had sunk within half a minute. The coxswain had to pick his way through the floating wreckage to pick up the survivors, three of whom were in a waterlogged boat. Two were on a small raft and one was lying across a couple of hatch covers. All seven were injured. Then on the 21st information came from the naval authorities at Brightlingsea that a ship had been blown up, and the lifeboat embarked doctors and put out to meet a sad procession of three naval drifters coming in with survivors of another drifter on board.

The *Edward Z. Dresden* was launched in eight minutes without waiting for a full crew on January 28th 1943, when Horlock's barge *Resolute* was mined. She found only wreckage, but took the skipper from another barge which had picked him up from the water.

Nearly sixty years after the *Albert Edward* capsized, disaster again struck the Clacton lifeboat, resulting in the loss of one of the crew. It was lunchtime on April 6th 1943, when the lifeboat was called out to the assistance of the barge *Tam o'Shanter*, which had a load of brick rubble for Maldon. When the lifeboat reached her she was anchored a mile and a half east of the Swin bell buoy. The skipper refused to leave the leaking barge and asked the coxswain to tow her into Brightlingsea, so second coxswain Frank Castle, assistant motor mechanic Arthur Liddiard and bowman A. Griggs were put on board the barge to help get the anchor up—the barge's crew consisted only of the skipper and a boy. Forty minutes later the lifeboat took the barge in tow, but after she had towed the *Tam o'Shanter* only a mile and a half the towline parted and it was decided to anchor for the night to await a more favourable tide.

During the night the wind and sea fell, and it seemed that the next day would be fine. In the morning, however, the wind increased again to an exceptionally strong gale from the west, with a heavy sea. Again the coxswain tried to persuade the skipper to abandon the 39-year-old barge, but he refused and suggested the lifeboat should tow her to Harwich in view of the changed conditions. The tide was against them, so the two vessels remained at anchor until one in the afternoon. The lifeboat then took the barge in tow, but after getting her through the Swin spitway the towline again parted, so the crew made sail on the barge and she went on her way, escorted by the lifeboat. Wind and sea were now almost abeam, and both lifeboat and barge were rolling heavily and shipping a good deal of water.

'What's all this water?' one of the lifeboatmen on board the barge asked as he heard water swilling about below. Suddenly the barge capsized, throwing the skipper, boy and the three lifeboatmen into the water. The lifeboat went at once to their aid, picking up Liddiard and Griggs; the crew managed to reach Frank Castle with the boathook but his oilie tore and they lost him—he seemed to be either dead or unconscious. Liddiard had been supporting the skipper of the barge, but before the lifeboat crew could get hold of him the latter had lost his hold and disappeared. The lifeboat searched for over an hour, but nothing could be found of the three who had disappeared.

Another barge with a cargo of brick rubble for Maldon was in trouble in the same area the following July. This was the Rochester barge *Maria*, which lost her sprit in rough weather and had to call for help. The *Edward Z. Dresden* was launched to go to her aid, but the engine broke down and the Walton lifeboat *E.M.E.D.* was called out to help. The Clacton crew rigged sails to get their boat back to station, however, and the *E.M.E.D.* was diverted by radio to the barge. She stood by through the night and next morning took her into Brightlingsea creek.

With the coming of peace the barbed wire and the anti-tank obstructions were cleared from the beaches and work went ahead on clearing defensive minefields, both at sea and on the shore. The waters that had seen so much violence during five years of war began to blossom with the white sails of yachts and small boats—and the lifeboat became busier than ever.

The first complete year of peace, 1946, which began with the lifeboat being called out to a tug disabled by an explosion on New Year's Day, provided a foretaste of what was to come. At 12.24 in the morning of February 1st came news of red flares which led to the *Edward Z. Dresden* rescuing three men from the auxiliary yacht *Puffin*, of Rochester, dragging her anchors half a mile off Jaywick beach after her engine had failed. Later in the day the lifeboat put out again to tow the yacht into Brightlingsea.

The rubber dinghies which had been the salvation of so many aircrew who had 'ditched' on their way back from operations during the war proved a curse when they were sold as 'government surplus' and came into the hands of foolhardy or careless holidaymakers. In May the lifeboat picked up a man, his wife and daughter who had been drifting in one off Jaywick for three hours and two men who were adrift in another in the same area.

The next month the *Edward Z. Dresden* towed the yacht *Pierrette* to Harwich and a couple of weeks later the reserve lifeboat *J. B. Proudfoot* towed another yacht, whose engine had broken down off Holland Sluice, into Harwich. On July 18th the *J. B. Proudfoot* took into Brightlingsea a converted ship's boat which had been blown off course when on passage from Southend for Rochford, and towed the old nine-ton racing cutter *Zem-Zem* into Burnham. Then, little more than a week later, she towed another yacht into Harwich, this time a Scots vessel whose rudder had been unshipped by heavy seas.

At the end of 1951 the *Edward Z. Dresden* was transferred to Stronsay, in the Orkneys, when that station was reopened after being closed for 37 years. After three years there the *Edward Z. Dresden* was replaced by a new boat and joined the reserve fleet. While acting as a reserve boat she herself suffered shipwreck, being driven ashore in Luce Bay, off the Mull of Galloway, as a result of a mechanical defect as she put into Portpatrick to refuel while on her way from Workington to Port Askaig. She survived and is still in 1968 in the reserve fleet.

Her replacement at Clacton was the 46 ft. 9 in. Watson cabin lifeboat *Sir Godfrey Baring*, named after the Institution's chairman, which had been on view at the Festival of Britain exhibition during 1951. In her first year of service the new boat carried out eight services, six to pleasure craft, and in her second year six services, of which five were also to pleasure craft.

The *Sir Godfrey Baring* was busy on February 1st 1953, when a north-westerly gale and a tidal surge created havoc all round the coast, the sea sweeping in to swamp low-lying parts of Essex. At 3 a.m. that day West Mersea police asked for the lifeboat to recover houseboats which had broken adrift from their moorings at the western end of Mersea Island, in the Blackwater. While the lifeboatmen were out securing these, the sea swept over the sea wall and inundated the shanty township of Jaywick, just to the south of Clacton, marooning some 600 people and drowning a number of others. Rescue work in commandeered and borrowed rowing boats went on throughout the day, but as the tide began to rise again there were still a number of families in danger in the flooded area, mostly cut off in the lofts of their wooden bungalows.

The possibility of dropping rubber dinghies from the air had been discussed, but this proved impossible and the police asked the lifeboatmen to see what they could do. The *Sir Godfrey Baring* was launched at 1.45 p.m. and, taking a dinghy in tow, the coxswain took her as close as possible to the sea wall. Then, by means of the dinghy, which the lifeboatmen had to manhandle over the sea wall and launch into the floodwaters on the other side, five men, a woman, two children, two dogs and a cat were rescued. They were taken back to the lifeboat and given hot soup and biscuits on the way back to Clacton pier.

The Clacton lifeboat has several times been called to the Blackwater to search for yachtsmen who were lost after falling overboard and to search for missing boats. Owing to the distance from its station and to the difficulty of manoeuvring a deep-draught lifeboat in shallow and congested waters these searches seldom resulted in life being saved, and thus West Mersea was one of the first places to receive one of the inflatable inshore rescue boats introduced by the R.N.L.I. to carry out holiday and other inshore lifesaving work. This carried out its first rescue operation at the end of July 1963, when it picked up two men whose dinghy had overturned in a fresh breeze. Since that time it has been called out more than 70 times. Like many other lifeboat stations, Clacton has since been allocated an I.R.B. in addition to its main lifeboat.

In 1967 it was announced that the *Sir Godfrey Baring* was to join the Institution's reserve fleet and that her place at Clacton was to be taken by a new boat of the 37 ft. Oakley self-righting type built by Herd and Mackenzie at Buckie in Scotland. Smaller than the previous boat, the Oakley would be able to launch at any state of the tide. Like her predecessor, she was given the name of a chairman of the R.N.L.I. committee of management, *Valentine Wyndham-Quin*. She was due to enter service early in 1968, but for some months was on the slip at Rowhedge Lower Yard while alterations were made to the pier slipway at Clacton to enable her to take over; the *Sir Godfrey Baring* was meantime based at Brightlingsea, as her predecessor had been at the beginning of the second World War.

BOAT RECORD, CLACTON

Years on station	Length, Breadth, Oars/Crew	Type, Weight, Cost	Year built, (Off. no.), Builder	Boat's name, Donor, Authority
1878- 1884	34′ 8′3″ 10 db/	SR £363	1878	*Albert Edward* United Grand Lodge of [Freemasons R.N.L.I.
1885- 1901	39′ 9′ 12 db/	SR £634	1884 (32) Forrestt, Limehouse	*Albert Edward* As previous boat R.N.L.I.
1901- 1929	45′ 12′ 6″ 12/	W 10t 17/ 14t 3 £1,890	1901 (463) Thames Iron Works, Blackwall	*Albert Edward*[1] As previous boats R.N.L.I.
1911- 1912	38′ 10′ 10/13	Liverpool 4t 3	1903 (516) Thames Iron Works, Blackwall	*Charles Deere James*[2] Leg. Mrs. S. C. Guthrie, [Tunbridge Wells R.N.L.I.
1929- 1952	45′ 6″ 12′ 6″ -/9	W cabin motor 20t 7 £8,496	1929 (707) Saunders, E. Cowes	*Edward Z. Dresden*[3] Leg. Edmond Dresden, [London R.N.L.I.
1952- 1968	46′ 9″ 12′ 9″ -/8	W cabin motor 21t 17 £24,628	1952 (887) White, Cowes	*Sir Godfrey Baring*[4] R.N.L.I. funds R.N.L.I.
1968-	37′ 11′ 6″ -/7	Oakley £37,000	1967 (985) Herd and Mackenzie, Buckie	*Valentine Wyndham-Quin* R.N.L.I. funds R.N.L.I.
Boarding Boat				
1901- 1934	31′ 7′ 3″ 10/-	ex-SR	1889 (255) Forrestt, Limehouse	*Ellen Newman and John Bentley*[5]

1 Motor fitted 1911. Transferred to Aranmore, 1929.
2 Reserve boat at Clacton while *Albert Edward* was having motor fitted.
3 Transferred to Stronsay, later joined reserve fleet.
4 Transferred to Wick, 1968.
5 From Newburgh.

SERVICE RECORD, CLACTON

Albert Edward Lifeboat				
1878	May	23	Brig *Garland*, of South Shields, saved	9
1879	Jan.	24	Ship *Hebe*, of Fredrikstad, saved	14
1881	Jan.	6	Barque *Hertha*, of Christiansand, gave help	
	Jan.	19	Barque *Hassels*, of Haugesund, saved	8
	Oct.	14	Schooner *Ocean*, of Goole, saved vessel and	4
	Oct.	23	Steam trawler *Madeleine*, of Boulogne, saved	16
1882	Apr.	29	Sloop *Nordstjernen*, of Haugesund, saved	5
1883	Mar.	30	Barque *Epsilon*, of Swansea, assisted to save vessel and	11
	Sept.	2	Barque *Rome*, of Arendal, saved	14
	Oct.	3	Barque *Danmark*, of Dragen, assisted to save vessel and	12
	Nov.	6	Brigantine *Estafette*, of Faversham, assisted to save vessel and	9
1884	Jan.	27	s.s. *Hawthorn*, of London, gave help	
Reserve Lifeboat				
	April	18	Brig *Thorley*, of Hartlepool, landed 8 from Swin Middle lightship	
	Sept.	7	Schooner *Mystery*, of Portsmouth, saved vessel and	5
Albert Edward (second) Lifeboat				
1885	Nov.	1	Barque *Garland*, of Aland, assisted to save vessel and	14
1886	April	7	Cutter yacht *Cruisada*, saved vessel	
	Nov.	5	Schooner *Ocean Bride*, of Guernsey, saved	9
1887	Oct.	10	Schooner *Venus*, of Teignmouth, stood by	
1888	Feb.	17	Brig. *Ipsilen*, of Krageroe, landed 9 from Swin Middle lightship	
	Feb.	21-22	Schooner *Dorothea*, of Riga, assisted to save vessel and	4
	Feb.	27-28	s.s. *Blonde*, of Whitehaven, assisted to save vessel and	10
	Apr.	25-27	Barque *Flekkefjord*, of Flekkefjord, assisted to save vessel and	12
Reserve Lifeboat				
	Nov.	5-7	Barque *Minnet*, of Helsingborg, saved	15
Albert Edward (second) Lifeboat				
1890	Oct.	3	Barque *Larissa*, of Blyth, picked up men from boats, saved	8
1891	Mar.	10	Schooner *J. W. Bebell*, of Beaumaris, landed a body and saved	1
	Oct.	7	Barque *Oliver Cromwell*, of Oland, saved	11
1892	Feb.	14	Barque *Saga*, of Flekkefjord, saved	11
	Feb.	16	Ketch *Don*, of Yarmouth, saved vessel and	6
1893	Nov.	20	Schooner *Nora*, of Mandal, saved vessel and	7
	Nov.	21	Barque *Harald Haerfagre*, of Fredrickstad, saved	11
	Nov.	22	Barque *Anna*, of Copenhagen, saved	10
1894	Jan.	4	Brigantine *St. Alexei*, of Copenhagen, saved	7
1895	Apr.	7	Schooner *Betty Russell*, of Lancaster, saved vessel	
	Nov.	30	s.s. *Vale*, of Stavanger, stood by	
1896	Nov.	8	Barquentine *Golgotha*, of Riga, saved vessel and	10
1897	Jan.	16	Brigantine *Pitho*, of London, stood by	
	Dec.	2	Barque *Thyra*, of Fredrikstad, saved	10
1898	Oct.	17	Barque *Inga*, of Laurvig, stood by	
	Nov.	22-23	Schooner *Ornan*, of Osthammer, saved	7
1899	Jan.	24	Schooner *Robert Anderson*, of Inverness, saved	5
	Jan.	26	Ship *Hawksdale*, of Liverpool, saved	18
			(Margate lifeboat saved another	7)
	Dec.	8	Schooner *Moonlight*, of St. Ives, saved	4
1900	Nov.	11	Barque *Norman*, of Laurvig, stood by	
	Dec.	18	Brigantine *Nina*, of Teignmouth, saved	6
1901	Mar.	22	Brigantine *Comforter*, of Tonsberg, saved	7
	Mar.	25	Smack *William and Eliza*, of Colchester, stood by	

Albert Edward (third) Lifeboat

1902	Feb.	2	s.s. *Dungonnel*, of Glasgow, assisted to save vessel and	11
	Mar.	15	Ketch *Carina*, of Brixham, saved vessel	
	July	1	Brigantine *Hilda*, of Abo, stood by	
1903	Sept.	9	Schooner *Martha*, of Rudkjobing, stood by	
	Oct.	14	Barque *Sheffield*, of Fredrikstad, saved	11
1906	Feb.	10	Yacht *Isidora*, of Dublin, saved vessel	
	Feb.	23	Schooner *Renner*, of Arbroath, saved	5

Boarding Boat

	Aug.	25	Yacht *Wild Rose*, of Burnham, saved boat and	4

Albert Edward (third) Lifeboat

	Dec.	1	Barge *Falcon*, of Maldon, saved	3
			Barge *Falcon*, of Maldon, saved vessel	
1907	Apr.	7	Ketch *Sabine*, of Trequier, stood by	
1909	Aug.	3	Barque *Magnat*, of Langesund, transferred 9 to tug	
1910	Jan.	14	H.M. Tug *Diligent*, assisted to save vessel	
	Feb.	12	Ketch *Cecilia*, of London, assisted to save vessel and	4
	Oct.	13	Lugger *Prince de Joinville*, of Boulogne, saved vessel and	21
	Dec.	1	Schooner *Pearl*, of Dundee, saved	5
	Dec.	4-8	s.s. *Antigone*, of London, assisted to save vessel and	24
	Dec.	9-11	s.s. *Antigone*, of London, gave help	
1911	Apr.	24	Yacht *Blackbird*, of Leigh, saved yacht and	2

Charles Deere James Reserve Lifeboat

	Nov.	13	Barge *Jumbo*, of London, stood by	

Albert Edward (third) Lifeboat

1912	Nov.	26	Motor boat *Konatzu*, saved boat and	2
1914	Jan.	11	Schooner *Mary Ann*, of Faversham, saved	6
	Apr.	1	Government seaplane *No. 11*, saved seaplane and landed	2
	Apr.	24	Government seaplane *No. 79* and motor launch, gave help	
	May	10	Yacht *White Heather*, of Southampton, saved yacht and	14
	June	4	Barge *Afternoon*, of Rochester, assisted to save vessel and	1
	July	19	Yacht *Doris*, of Burnham, saved vessel and	5
	Dec.	2-5	s.s. *Harlington*, of London, saved	15
	Dec.	12	Barge *Lorna*, of Greys, assisted to save vessel	
	Dec.	29	Barge *Five Brothers*, of Rochester, assisted to save vessel	
	Dec.	30	Barge *Renown*, of London, saved vessel and	2
1915	Jan.	17	Minesweeper H.M.S. *Eaton*, stood by	
	Apr.	8	Barge *Mary Ann*, of Harwich, stood by	
	Sept.	20	s.s. *Capitol*, of London, stood by	
	Sept.	28-29	Barque *Leading Chief*, of Guernsey, saved	9
1916	May	9	Barge *Evening*, of Rochester, saved vessel and	2
	June	3	Government seaplane, saved seaplane and	3
	June	4-7	s.s. *Corliz*, of Bilbao, assisted to save vessel and	24
1917	Dec.	27-28	s.s. *Iris*, of Helsingborg, saved	23
1919	Apr.	26	Barge *H. Pierrepont*, of Rochester, saved	2
	June	21	Yacht *Bonita*, of Burnham, assisted to save vessel	
	Dec.	23	s.s. *Stanley Hall*, of Leith, assisted to save vessel and	17
1920	Jan.	8	H.M. Tug *Dromedary*, saved	18
	Oct.	1	Rowing boat *Ellen*, of Clacton, saved boat and	3
	Dec.	16	s.s. *Limehouse*, of London, stood by	
	Dec.	18-21	s.s. *Limehouse*, of London, gave help	
1921	Sept.	21	Russian barquentine *Esther*, stood by	
	Oct.	29	Barge *Mary Graham*, of Rochester, saved vessel and	3
	Nov.	7	s.s. *Baltic*, of Christiania, stood by	
	Dec.	24	s.s. *Sheaf Arrow*, of Newcastle, stood by	

1922	Apr.	17	Cutter *Dulcebella*, of Westcliff, saved vessel and	6
	Aug.	7	Cutter *The Wasp*, of Rowhedge, saved vessel and	11
1923	Jan.	20	Barge *Decima*, of London, saved vessel and	2
	May	13	Yacht *Irene*, of Rochester, saved vessel and	2
1924	Feb.	16	Schooner *Alatga*, of Tvedestrand, stood by	
	July	27	Yacht *Elidor*, of Burnham, saved vessel and	3
1925	Feb.	11	Ketch *Rosina*, of Plymouth, saved	4
	May	17	Yacht *Chione*, of Southampton, stood by	
	July	29	Schooner *Walkyrie*, of Lannion, saved vessel and	1
	Sept.	26	Yacht *Dulcebella*, of Westcliff, saved vessel and	2
1927	July	23	Steam tug *Sun IX*, of London, stood by	
	Aug.	22	Yacht *Moira*, of Rochford, saved vessel and	2
			A boat of Whitstable, saved boat and	1
	Dec.	26	s.s. *Fernwood*, of London, stood by	
1928	Apr.	3	Yacht *Coquette*, of London, saved vessel and	4
	Aug.	27	Yacht *Iris*, of Colchester, saved vessel	

Edward Z. Dresden Lifeboat

1929	Aug.	6	Yacht *Linnet*, of Burnham, saved vessel and	2
	Sept.	3	Motor launch *Princess Olga*, of Burnham, saved vessel and	3
1930	Feb.	16	Cutter *Elizabeth*, of Colchester, saved	2
	June	15	A motor boat, saved boat and	3
	Dec.	11	Barge *Squeak*, of London, assisted to save vessel and	2
1931	Aug.	8	Motor boat *Grace Darling*, of Holland-on-Sea, landed 2	
1932	Sept.	11	Two small boats, saved boats and	3
1933	Apr.	9	Yacht *Amy*, of West Mersea, saved vessel and	5
	July	31	Barge *Audrey*, of London, saved vessel and	2
1934	Aug.	11	P.s. *Clacton Queen*, of Rochester, gave help	
	Dec.	18	Barge *William Cleverly*, of Rochester, stood by	
1935	Jan.	27	Barge *T.T.H.*, of London, assisted to save vessel and	3
	Apr.	10	Barge *Arthur Margetts*, of Rochester, saved vessel and	1
1936	Apr.	7	Barge *Edith and Hilda*, of Rochester, stood by	
	July	18	Barge *Victa*, of London, saved vessel and	2
	Aug.	3	Small boat of Jaywick, saved boat	
	Oct.	5	Smack *Arco*, of Leigh, saved boat and	6
	Nov.	2	Rowing boat, saved boat	
	Dec.	8	m.v. *Conida*, of London, gave help	
1938	May	4	Barge *Florence Myall*, of Rochester, saved	2
	May	30	Yacht *Westwind*, of Bremen, gave help	
	June	2	Yacht *Das Wappen von Bremen*, of Bremen, saved vessel and	7
1939	Jan.	3	Barge *Louise*, of Rochester, assisted to save vessel and	2
	Apr.	22	Yacht *Sprite*, of Leigh, saved vessel and	2
	June	15	Motor yacht *Mayflower*, of Yarmouth, saved vessel and	2
	June	28	Motor yacht *Alanfred*, of Benfleet, saved vessel and	4
	July	15	Aux. yacht, *Marosa*, of West Mersea, stood by	
	Oct.	25	Rowing boat *Mon Ami*, saved boat and	3
	Nov.	19	Barge *Lorna*, of London, saved	2
			Aux. smack *Charlotte Ellen*, of Tollesbury, saved	3
	Nov.	24	s.s. *Lightfoot*, of Newcastle, stood by	
	Nov.	25	Aux. barge *Lady Dorothy*, of Ipswich, stood by	
	Dec.	1	s.s. *Perth*, of Dundee, stood by	
1940	May	30	Station temporarily closed	
	May 31 - June 4		Lifeboat took part in Dunkirk evacuation with a naval crew.	
	August		Station re-opened at Brightlingsea.	
	Nov.	7	H.M.S. *Reed*, gave help	
	Nov.	8	s.s. *Fireglow*, of London, escorted vessel	
	Nov.	13	Station re-opened at Clacton	
	Nov.	18	m.v. *Ability*, of London, saved	7
	Nov.	21	Royal Navy patrol drifter, gave help	

J. B. Proudfoot Reserve Lifeboat

1941	June	25	Aircraft, salved dinghy and saved	1

Edward Z. Dresden Lifeboat

1941	Sept.	16	Aircraft, gave help	
	Nov.	17	Barge *Saltcote Belle*, stood by	
1942	Jan	13	Barge *Glenmore*, of Rochester, saved vessel and	2

The Brothers Reserve Lifeboat

1942	Apr.	2	Dinghy of Jaywick, saved boat and	1
1943	Jan	28	Barge *Resolute*, of Mistley, landed 1	

Edward Z. Dresden Lifeboat

1945	Jan.	13	Royal Navy minesweeper, landed 2, saved	11
			Aircraft, saved	1
	Jan.	19	Barge *Thalatta*, of Harwich, saved vessel and	3
	Dec.	15	*M.F.V. 125*, gave help	
1946	Feb.	1	Aux. yacht *Puffin*, of Rochester, saved vessel and	3
	May	23	Two rubber dinghies, saved dinghies and	5
	June	10	Aux. yacht *Pierette*, saved vessel and	4

J. B. Proudfoot Reserve Lifeboat

1946	June	27	Aux. yacht *Doris*, of Shoreham, gave help	
	July	18	Converted ship's boat *Harriette*, of Rochford, saved boat and	3
			Racing yacht *Zem Zem*, of London, saved boat and	2
	July	30	Yacht *Coya*, of Greenock, saved vessel and	4

Edward Z. Dresden Lifeboat

1948	Jan.	30	Motor barge *Arripay*, of London, saved a boat and	3
	Aug.	9	Aux. yacht *Sea Air*, of Southend, salved a dinghy	
	Sept.	13	Barge *British Oak*, of London, saved	2
	Dec.	6	Barge *Esterel*, of London, saved	3
1949	July	5	Canoe, gave help	
1950	Apr.	12	Yacht *Dirk II*, of Maldon, stood by	
	July	16	Yacht *Myrtle*, of Falmouth, escorted yacht	
			Yacht *Morning Mist*, of London, gave help	
	Aug.	24-25	Motor barge *Lewes Castle*, of London, landed 3 and gave help	
1951	Mar.	23	Yacht *Buxey*, of Maldon, gave help	
	May	27	Yacht *St. Helier*, of Burnham, gave help	
	June	23	Yacht *Spray*, of Woodbridge, gave help	
	Aug.	5	Yacht *Serento*, of St. Osyth, gave help	
	Aug.	7	Yacht *St. Helier*, of Maldon, landed 2 and gave help	
	Aug.	12	Yacht *Christine*, of Burnham, saved vessel and	2
	Aug.	26	Yacht *Ingrid II*, saved vessel and	2
	Aug.	27	Motor cruiser *Thelas*, of Hull, stood by	
	Sept.	25	Yacht *Ayesha*, of Brightlingsea, saved vessel and	3
	Oct.	20	Motor cruiser *Widgeon*, of Ramsgate, gave help	
	Dec.	9	Smack *Iris*, of Colchester, saved boat	

Sir Godfrey Baring Lifeboat

1952	Feb.	4	Motor barge *Delce*, of London, gave help	
	May	28	Motor cruiser *Susie*, of Maldon, saved vessel and	2
	June	1	Barge yacht *De Bries*, gave help	
	July	23	Rowing boat, landed 1	
	Aug.	5	Yacht *Judy*, gave help	
	Aug.	9	Motor boat *Jill*, of Clacton, saved boat	
	Sept.	25	Yacht *Airbourne*, of London, saved vessel and	2
	Oct.	19	Motor fishing vessel *Monbretia*, of Scarborough, saved boat and	4

1953	Feb.	1	Persons marooned on bungalow roof at Jaywick, saved	8
	Apr.	1	Motor cruiser *Frances N. Dee*, gave help	
	June	21	Yacht *Dottie*, of Holland-on-Sea, saved yacht	
	July	7	Yacht *Daddy*, of Maldon, saved boat and	2
	Aug.	23	Barge yacht *Dobber*, of Bradwell, saved yacht and	5
	Sept.	3	Yacht *Cloetta*, of Burnham, gave help	
1954	June	9	Motor boat *Nemo II*, of Clacton, saved boat	
	July	22	Motor cruiser *Holliwell*, gave help	
	Dec.	22	Rowing boat, saved boat and	2
			Man fallen from pier, saved	1
1955	Feb.	20	Trawler *Denise Germaine*, of Zeebrugge, saved a boat	
	Feb.	21	Barge *May*, of Ipswich, saved vessel and	2
	Mar.	19	Motor barge *Persevere*, of Rochester, gave help	
	July	30	Yacht *Ayesha*, gave help	
	Aug.	6	Motor yacht *White Heather*, saved yacht and	1
	Sept.	24	Yacht *Johnnie Belinda*, of Wallasey, gave help	
1956	Apr.	14	Yacht *Sandling*, of Norwich, gave help	
	May	24	Motor tanker *C653*, of London, stood by	
	May	27	Schooner *Lora*, of East Mersea, escorted vessel	
			Fishing vessel *HH100* of Harwich, saved boat and	4
	June	3	Sailing dinghy, saved dinghy	
	July	29	m.v. *Sunny Seas*, of Bradwell, gave help	
	Aug.	1	Yacht *Drofli*, of London, saved	4
	Aug.	19	Yacht, saved yacht and	1
	Sept.	27	Aux. barge yacht *Winifred*, of Colchester, saved	5
	Oct.	6	Fishing boat *Melona*, saved boat and	2

H.C.J. Reserve Lifeboat

1957	June	30	Two swimmers marooned on sandbank, rescued	2
	July	7	Yacht *Peace*, of Colchester, saved yacht and	3
	Aug.	18	Sailing dinghy, gave help	
	Aug.	24	Yacht *Mariet*, of Maldon, saved boat and	3

Sir Godfrey Baring Lifeboat

1957	Oct.	20	Yacht *Starbeam II*, saved boat	
1958	June	19	Sailing yacht, saved boat and	3
	July	14	Yacht *Freedom*, gave help	
	Sept.	8	Yacht *Trojan*, gave help	
	Oct.	3	Yacht *Stella*, of Whitstable, saved	1
	Oct.	16	Barge *Alan*, of London, stood by	
1959	Feb.	21	Motor yacht *Herga*, of St. Helier, gave help	
	July	11	Catamaran, saved boat and	1
			Yacht *Kestrel*, saved boat and	4
	July	12	Dutch m.v. *Bermuda*, stood by	
	July	27	Rowing boat, saved boat and	1
	Aug.	1	Sailing dinghy, saved	1
1960	Apr.	9	Ex-ship's boat *Tot Morgen*, gave help	
	May	8	Motor cruiser *Cheviot Wind*, of Yarmouth, gave help	
	May	18	m.v. *Watchful*, of London, stood by	
	July	9	Yacht, landed 2	
	July	10	Yacht, gave help	
	Sept.	3	Yacht *Jetheu*, escorted vessel	
	Sept.	12	Motor cruiser *Vacation*, of Maldon, stood by	
	Nov.	19	*M.F.V. 1256* from H.M.S. *Ganges*, escorted boat	
1961	May	21	Motor cruiser *Pesaro II*, saved boat and	2
	Aug.	3	Yacht *Cramarye*, gave help	
	Aug.	4	Yacht *Janet*, of Maldon, gave help and landed 2	
	Aug.	5	Yacht *Spray*, gave help	
	Aug.	8	Man in sea, rescued	1
	Aug.	14	Motor boat, saved boat	

1962	Sept.	7	Capsized catamaran and motor boat, escorted boats	
	Nov.	4	Rowing boat, saved	2
	Nov.	14	Polish m.v. *Krutynia*, stood by	
1962	Mar.	7	Barge *Resourceful*, of Manningtree, saved	2
	Apr.	17	Ex-police launch *Marina*, gave help	
	May	12	Motor cruiser *Fluminence*, gave help	
	May	31	Motor cruiser *Amouelle*, of London, saved boat and	2
	June	26	Canoe, saved canoe	
	Aug.	10	Sailing boat *Oberon*, of Brightlingsea, saved	1
	Aug.	11	Yacht *Volma*, saved boat and	3
1962	Aug.	23	Yacht *Aslaug*, saved boat and	4
	Oct.	11	Motor boat, saved boat and	3
	Oct.	28	Motor cruiser *Buahari*, saved boat and	5
1963	Jan.	21	24 motor vessels in Blackwater, gave help to three vessels and landed 22	
	June	9	m.v. *Viking*, escorted vessel	
	June	24	Yacht *White Heather*, gave help	
	Aug.	25	Six dinghies, gave help	
	Sept.	8	Yacht *Davneen*, gave help	
	Sept.	24	Yacht *Jan Z. Kolna*, of Gdynia, stood by	
	Oct.	27	Motor cruiser *Gay Bet*, of Brightlingsea, gave help	
	Nov.	17	m.v. *Heiloo*, of Flushing, stood by	
1964	Apr.	5	Yacht *Suzy of Chalfont*, gave help	
	Apr.	29	Rowing boat, saved boat and	2
	June	7	Smack *MN41*, of Maldon, gave help	
	Sept.	15	Sloop, saved	2
	Oct.	16	Rowing dinghy, landed 3	
	Nov.	13	Yacht, stood by	
1965	Jan.	17	Rowing boat, saved	4
	July	11	Catamaran, saved boat and	2
			Yacht *Benito Elinor*, stood by	
	July	25	Yacht *Nassau Girl*, stood by	
	Aug.	1	Speedboat, gave help	
	Oct.	2	Motor cruiser *Seabird* and	
			motor cruiser *Smuggler II*, saved boats and	2
	Nov.	27	Motor barge *Rodent*, stood by	
1966	Jan.	12	Radio ship *Radio London*, stood by	
	Mar.	27	Yachts, *Essex Pride*, *Folk Dance*, *Lion Heart*,	
			Aslaug and *Contessa*, saved boat and rescued	5
			and gave help to four other boats	
	May	23	Yacht, saved boat	
	Aug.	14	Yacht, stood by	
	Oct.	25	Rowing boat, saved boat and	2

6

SOUTHEND

SOUTHEND, known today as a seaside resort with the longest pier in the world, lies near the point at which the Thames estuary narrows abruptly to a width of about four miles. Between the Essex shore and the Isle of Grain on the Kent shore lie narrow deep-water channels used by shipping making for the port of London; along the Essex coast lies the Chapman sand and to the east beyond Shoeburyness lies the Maplin sand, while the Kent coast is bordered by the Yantlet flats, and in between is the Nore sand. In these narrow waters small craft and some of the biggest ocean-going ships are constantly passing—and, in times gone by at least, frequently colliding.

Prittlewell was little known in the early years of the nineteenth century, but it became popular after Queen Caroline and Princess Charlotte had paid a visit in 1804. The south end of Prittlewell, soon to become known simply as Southend, became a popular resort during the course of the century, the first pier being built between 1828 and 1830. The pierhead was at first a floating hulk and later a piled structure known as The Mount, separated from the pier itself, from which passengers were transferred by rowing boat. The gap was closed in 1846.

This wooden pier was still in existence when the R.N.L.I. sent the first lifeboat to the town in 1879. Paid for by a special fund collected by Edwin Brett through the magazine *Boys of England*, this boat was only 25 ft. long and 7 ft. 3 in. in beam, pulling eight oars double-banked—little more than a ship's boat, and in fact launched from davits on the pier with the help of two small winches. Although described as a pulling and sailing boat, the *Boys of England and Edwin J. Brett* appears in fact to have been intended primarily for rowing, and her two lugsails were hardly ever used. It was thought at this time that such a craft was sufficient for work in the Thames estuary, a comparatively sheltered area in which at that time more casualties occurred through collision than through stress of weather.

The first coxswain of the Southend lifeboat, W. C. Bradley, was quite a colourful character. At the time he became coxswain he was in charge of the pierhead, and it was because of his employers' expressed desire that he should not leave the pierhead that he had to relinquish the post of coxswain after only a relatively short time. They apparently did not wish him to leave the pierhead under any circumstances; he held his appointment for twenty years and had no holiday during that period. He left the service of the Local Board in 1891 and was elected to the town council ten years later, eventually becoming an alderman,

The *Boys of England* performed her first service slightly less than two weeks after the naming ceremony, going out on November 26th 1879 to the help of the *Wilhelmina*, of Copenhagen, which had grounded on the Nore sand. Her assistance was not needed, however, for a tug and three smacks had already been engaged.

The following March the lifeboat was called out twice in one day, the second time to a barge in danger in the south channel near the Jenkin sand. The barge, the *William Little*, of Rochester, had drifted from her anchorage under the Yantlet flats and had capsized close to the Jenkin swatchway. The lifeboat could make little headway against the wind and a strong tide, so four Leigh fishermen and two coastguard boatmen manned a large boat, the *Three Sisters*, belonging to one of the Leigh men and picked up the barge's crew.

In 1885 it was decided by the R.N.L.I. to send a second lifeboat to Southend, a larger one which would be kept in a boathouse. This was the *Theodore and Herbert*, built at a cost of £300—little more than a third of the cost of the boat which replaced her and less than one hundredth that of the present Southend lifeboat. The lifeboat-house built at this time at a total cost of £500 is still standing at the bottom of Hartington Road, though like much else in Southend it has turned its back on the sea and is now used in connection with the holiday industry.

In the 1880's, however, Southend retained much of the character of a fishing village, and some of the pleasure boats which operated from the beach during the short summer season had temporary decks fitted in winter so they could go stowboating. Some of these pleasure boats were of no mean size; in the early 1880's the largest, Joe Myall's *Conqueror*, was about 40 ft. long from stem to counter stern, but when she was run down by a steamer in the Princes channel while on a spratting trip she was replaced by a much larger boat of the same name. Largest of all were the *Prince of Wales* and the *Victoria*, yacht-like craft with bright varnished hulls and spotless white canvas. The *Victoria* was licensed for some 300 passengers and the *Prince of Wales* for about 240.

There were lifeboats and 'lifeboats' at Southend in those days, for about 1890 George Murrell bought an old beach yawl from Norfolk, replaced her lugs with gaff sails and painted 'Murrell's Lifeboat' on her nameboard. She proved most successful as a pleasure-tripping boat, as did other former beach yawls which went into the same business. So successful was she, in fact, that Murrell had several boats built locally to a similar design, though these were carvel built instead of clinker. The 'lifeboats' made only a short trip and did not attempt to compete with the bigger craft, which took trippers round the Nore lightship or across to Sheerness.

Nearby Leigh was a busy centre of the fishing trade in those days, and towards the end of the century there were about a hundred boats engaged in shrimping. Fortunes fluctuated wildly and when things were bad in the Thames estuary the Leigh fishermen wandered far afield, to Harwich and to the French coast. Casualties were frequent and hardly a year went by without the sinking of a boat or the loss of several lives, so it was not surprising that in 1866 an insurance club for the fishing boats was formed. This club bought a large boat for use in case of accidents in 1887. She was named *The Lifeguard*, but details of any lifesaving work are unknown.

When the barge *Maud Hawthorn* sank off Shoeburyness coastguard station, just to the west of the ness itself, in an exceptionally heavy gale on March 11th 1888, it was neither of the two lifeboats which was responsible for the rescue of the crew. The coastguardsmen made four attempts to launch their own boat, but they could not get her off the beach and she was stove in. So the news was taken to Halfway House, a public house which is still in existence mid-way along the front just where the coast begins to curve south again, and some local watermen launched a boat, the *Victoria*, in which they succeeded in reaching the wreck. They had great difficulty in getting the mate out of the rigging in which he had hung for several hours, but they managed it in the end.

Such gales as this created havoc among the considerable fleets of barges then using the Thames. On this occasion the barge *Edward*, laden with coal for Colchester, 'drew her anchor' and drove on to Vandervord's pier at Southend, doing considerable damage, whilst another barge, the *Two Sisters*, also dragged ashore. The *Edward* was partly unloaded and got off on the night tide, when the storm had abated somewhat.

A second *Boys of England* was sent to Southend towards the end of

The Clacton motor lifeboat Edward Z. Dresden, *built in 1929, is still in the Institutions reserve fleet. She is a 45 ft. 6 in. Watson-type boat.* *(Photo: R.N.L.I.)*

The crew of the third Clacton lifeboat Albert Edward *with survivors of the schooner* Pearl, *rescued from the wreck of their vessel on the Blacktail spit in December 1910.* *(Photo: J. Theobald)*

Southend's first lifeboat, the 25ft. pulling boat Boys of England and Edwin J. Brett, *on Southend beach about 1879—a rare photograph. In the background is the old wooden pier.* *(Photo: Southend Public Library)*

The Southend lifeboat Greater London (Civil Service No. 3) *searching for survivors from the submarine* Truculent, *sunk in collision on January 12th 1950. (Photo: R.N.L.I.)*

1889. Formerly at Weston-super-mare—where, as at Southend, she was kept in davits on the pier—she seems to have been similar in all respects to her namesake and predecessor. She had only a short reign at Southend, for she was sent to Southport as boarding boat in 1891 and the No. 1 station was closed.

With the closing of the No. 1 station it was decided that the *Theodore and Herbert* should be moored inside the pierhead during the winter months so as to be more readily available for service.

The old wooden pier had at this time just been replaced by a new iron structure, and the horse tram which had done duty since the 1840's was replaced by an electric one, then a considerable novelty. This tramway has often proved a blessing to the lifeboatmen. Nowadays a train driver is always on standby duty and turns out whenever the lifeboat is needed, but in earlier days—until 1915, in fact—the tramway was available only between 6 a.m. and midnight, and if a call came in the early morning the crew faced a long tramp to the pierhead.

In the twelve years the Southend No. 1 station was in existence the boats were launched a dozen times and saved 23 lives. The *Theodore and Herbert* was transferred to Workington, Cumberland, in 1899 and replaced by a 38 ft. boat, the *James Stevens No. 9*, one of several provided from a legacy of James Stevens, of Birmingham. This boat remained longer and did much more work than her predecessors, for she lasted 24 years and was to be launched on service no fewer than 111 times, saving 94 lives.

A boat of the Norfolk and Suffolk type—heavy and powerful, depending mainly on sails for propulsion and with water ballast confined in tanks coupled with a broad beam to give her a high degree of stability—the *James Stevens No. 9* proved much more useful than her self-righting predecessors. There was plenty for her to do in the busy waters of the Thames estuary; the ocean-going liners and tramp steamers that passed in daily procession only a short distance from her station at the pierhead were generally able to look after themselves and keep out of trouble, but barges and other small craft were frequently in distress, particularly when foul weather brought up a lumpy sea in the narrowing estuary.

The story of the spritty *The Sisters*, of Ipswich, is one which was repeated over and over again. She left Felixstowe Dock for Rochester on January 31st 1913, with barley screenings. It took her six days to reach Burnham, and after a night in the Crouch she made a further effort to get across the main estuary. Then a south-westerly gale sprang up and before sail could be shortened the topsail and jib blew out, the foresheet parted and the foresail blew to pieces. The crew hoped they could make the East Swale, on the Kent coast, under main and mizzen, but then the steering chain broke. They anchored the crippled barge inside the West Oaze buoy, but before help came she sank by the head, the skipper and mate being washed off the deck as she went down. Fortunately the small boat floated clear and they were able to haul themselves into it, eventually reaching the Mouse lightship, the mate rowing on the lee side and the master using the other oar in the sculling notch in the transom. They were in due course landed from the lightship by the Southend lifeboat.

In one day, on December 29th 1914, the *James Stevens* performed three services, the first of them to a London tug, the *Woodcock*, which the lifeboatmen found drifting broadside to the sea. After three attempts they succeeded in getting alongside. They found her in charge of the mate—the skipper had gone ashore—and without sufficient steam to get way on her. The lifeboatmen went on board, cleared the anchor which was foul, and brought the tug up. The pumps, they then discovered, were choked, so they bailed her out with buckets in order to reach the bunkers. When steam was raised they brought the tug in and moored her inside the pierhead.

After completing this job the signalman reported that a barge was in difficulties on the West Shoebury sand. The lifeboatmen waited only to put ashore one of the crew who was ill and to take on board a replacement before putting off to the aid of the barge, the *Eliza*, of Rochester. The lifeboat took off the barge's crew of two, and then four of the lifeboatmen volunteered to board the casualty and take her into Shoebury bay.

The third call was to the barge *John Tinnoth*, of Rochester, a spritty built at Frindsbury in 1851, which was foul of the steamer *Uno*, anchored in the channel about a mile and a half west of the pier. The skipper was glad to have help, so four lifeboatmen were put on board the barge to help get her to Southend.

The lifeboatmen's help was not always welcome, however. Only a few months later the second coxswain noticed a barge stranded on the Grain Spit, with seas washing over her. The wind was dead against the lifeboat and the tide was flooding, so the lifeboatmen had to obtain the help of the tug *Mosella* to get away from the pier. They found the barge *Louise Lorne*, a London-registered spritty built at Nine Elms in 1871, being assisted by the Southend motor boat *Britannia*. Their own

The Mount, the head of Southend's old wooden pier, as it looked at about the time the pulling and sailing lifeboat Boys of England *was sent to the station in 1879. This pier, with its horse tramway, was replaced in 1887-89 by an iron structure which has since been extended several times.*

offer of help was refused, but the lifeboat stood by until the barge sank; the lifeboatmen said afterwards that had advantage been taken of their offer the barge would probably have been saved.

The *James Stevens No. 9* was condemned as unfit for further service late in 1923 and a temporary boat, the *William Roberts*, a 41 ft. Watson boat which had been stationed at Littlehaven, Pembrokeshire, until that station closed in 1921, was sent to Southend. During a brief stay she responded only to two fruitless calls, being withdrawn in June 1924 on the arrival of the *Charlie Medland*, a 43 ft. Watson boat which had been built in 1904 for The Mumbles. During her 20 years at this Welsh station she had saved 45 lives, and at Southend she added another seven to her record.

Though the *Charlie Medland* performed eight services during her four years at Southend all seven people saved were rescued on two days in 1925—two from a barge on January 31st and three from a Brixham cutter and two from a barge on April 16th.

In May 1928, a new motor lifeboat, one of the three 'Ramsgate' type, was sent to the station. Built by J. Samuel White at Cowes, the *Greater London* (*Civil Service No. 3*) was presented by the Civil Service Lifeboat Fund to commemorate the service of William Fortescue Barratt, who had been honorary secretary of the fund from 1914 to 1928.

Sometimes it was the bigger ships which were the cause of the lifeboat being called out. On October 6th 1928, the Ellerman liner *City of Lancaster* radioed that she was in distress in the vicinity of the Nore lightship. The lifeboatmen found her aground on the west side of the Oaze sand, badly holed on the starboard side and with her after holds flooded. 'Offered assistance which was declined without thanks; stood by until three tugs arrived; then returned to station,' stated the report.

But it is interesting to note that more than a third of the services carried out by the *Greater London* were to sailing or motor barges. One of her busy days was November 23rd 1930, when several of these, lying near the Jenkin sands off the Isle of Grain, got into difficulties. The *Charles and Esther*, from Great Wakering, the *Marion*, a London and Rochester Trading Co. barge, and the *Emily*, one of R. & W. Paul's barges from Ipswich, dragged their anchors and became dangerously waterlogged. The crews burnt clothing and bedding soaked in paraffin to attract attention, and these signals were seen by the berthing master on Southend pierhead, who called the lifeboat crew at 12.40

a.m. The *Greater London* went from one barge to another taking off the crews, who had to leap from barge's deck to lifeboat as the *Greater London* was run up alongside, Coxswain Sam Thomas having to put about and return three times before the skipper and mate of one of the barges could get on board.

Later in the morning the three crews were taken out to the barges. The *Charles and Esther*, which was loaded with wheat, had sunk and only her topmast was visible, but the crews of the other two were put on board their craft. The *Charles and Esther* was raised the following week by the Port of London Authority and towed up the Thames.

Nearly fourteen months later the *Greater London* had another busy day. The motor barge *Cader Idris* left Sea Reach on January 5th 1932, towing another former spritty, the *Stanley Margetts*. The weather blew up rough as they made their way down the estuary towards Oysterfleet on Canvey Island, and from 7 p.m. the skipper and his son on the motor barge were pumping continuously. In spite of all their efforts the *Cader Idris* began to sink by the stern as the crew sent up flares, setting fire to petrol-soaked canvas and ropes in a desperate attempt to attract attention. They scrambled on board the *Stanley Margetts* just before their own barge went down, but their relief was short lived, for the *Stanley Margetts* was also sinking fast, and when they were found by the lifeboat near the Low-way buoy, about midway between the Chapman light and Leigh, they lost no time in getting to the safety of the lifeboat.

As they were landing the three rescued men at the pierhead the lifeboatmen saw distress signals from the Nore. So out they went again, first to the lightship and then to the Mouse lightvessel, where they were told that flares had been seen from further out. They found the spritty *Lady Mary* high and dry on the sands, and stood by until she refloated safely as the tide flowed.

A little more than a week later, on January 14th, another barge was in trouble as the wind suddenly strengthened in the early hours of the morning. The Rochester spritty *Conqueror* had just passed the pier on her way to London with a cargo of sand when the head of the topmast broke off. Then the foresail blew out and the covering of the forehold burst as waves swept across the decks of the heavily laden barge. To make matters worse the steering gear broke and the vessel became unmanageable. The lifeboat came up with the barge a few hundred yards west of the pier near the edge of the deep water and took the skipper and the 17-year-old mate out of her. The *Conqueror* kept afloat, however, and the crew were able to return to her later the same day.

Ever since 1891 the lifeboat had been kept afloat inside the pierhead, at least throughout the winter. In 1935, however, a new boathouse and slipway was constructed at the end of the pier, costing £15,750 compared with the total of less than £500 for the old lifeboat-house ashore.

Fog in the approaches to a large port such as London can be as great a hazard as heavy weather. The fog must have been more than a little irritating to the Southend holidaymakers in May 1937 when it hid both the arrival and the departure of the Home Fleet, which spent six days at anchor off the town. But fog in the Thames was much more than just an annoyance to holidaymakers on the beach. It was fog blanketing the estuary which caused a collision between the London and Edinburgh Shipping Company's steamer *Royal Archer* and the Glen Line vessel *Gleniffer*, which was at anchor off the South Shoebury buoy, on the evening of June 12th 1937. The *Royal Archer*, which had just begun a voyage from London to Leith with 120 passengers, was badly holed amidships and water poured through the shattered plates into the engine room. The master decided to beach her, and the passengers were summoned from dinner to don lifejackets and prepare to leave the ship. His decision avoided disaster, and the passengers were able to file quietly down the accommodation ladders into pleasure boats from Southend which joined the reserve lifeboat *J. B. Proudfoot*, then at Southend, and an Admiralty tug from Sheerness in taking off the passengers and their baggage.

The lifeboat took 49 passengers to the pierhead and then stood by the *Royal Archer* until the early hours of the next morning. The fog had cleared when the P.L.A. wreck barge *Yantlet* and four tugs from Gravesend got the steamer off after a collision mat of wooden battens and tarpaulins had been fastened over the gap in her plating, and crowds thronging the pier and the sea front at Southend watched as the tugs towed her up the estuary towards Gravesend.

Curiously, both vessels had been involved in collisions in the Thames before. The *Gleniffer*, a 9,429-ton steamer built by Hawthorn, Leslie on the Tyne in 1915 and employed in trading with the Far East, had been in collision off Gravesend with the Russian vessel *Kamenetz Podolsk* in February 1936 and the *Royal Archer*, a 2,266-ton vessel built by Scott's of Greenock in 1928, had been in collision with the Donaldson liner *Cordillera* in Gallions Reach in May 1933.

Coxswain Sidney Page, who was a lifeboatman for 44 years and coxswain for just over 21 of these, was awarded the Bronze Medal for a series of dangerous and difficult rescues on June 1st 1938. The lifeboat was called first to the yacht *Wimpie*, which was in temporary difficulties off the pierhead.

Just as the lifeboat crew returned from this mission they saw a flare from the Shoebury sands, where the motor barge *Glenrosa*, bound from Rowhedge to London with sand, had blown aground after the engine had failed and the anchor chain had parted. When the crew of three tried to launch the barge's boat the painter snapped and the boat was swept away. They just had time to light a flare before the barge filled and sank and they were forced to climb into the rigging, from which they were taken by the lifeboat.

Then at 4.30 a.m. came a message that another barge was in trouble in the Low-way, off Leigh. Out went the lifeboatmen again, to find the *Maid of Monaghan* being lashed by heavy seas. The crew refused assistance, but the lifeboat stood by until rockets were seen from the Nore lightship. Reaching the lightship, they were told of a barge drifting helplessly, her sails in ribbons, This was the *Audrey*, owned by E. J. and W. Goldsmith Ltd., which was in desperate straits when the lifeboat arrived, so the *Greater London* took off the crew of two and their dog and then went to search for the barge *Myra*, which had been sailing with the *Audrey*. Ultimately the lifeboatmen found her mast sticking up three feet above the water and it was feared that the crew had drowned although the skipper and mate were found later at Bradwell after drifting for over five hours in the barge's small boat.

On this same occasion, at Southend, the locally owned barge *Portage* crashed into the Camper Road jetty, her mast snapping off and falling across the jetty promenade. Immediately to the eastward another locally owned barge, the *Haste Away*, also crashed into the jetty, and to the westward a third locally owned spritty, the *Ashingdon*, was thrown on to the sand against the sea wall. Goldsmith's barge *Viper* was badly damaged by being battered against the gasworks jetty, and the barge *Eureka*, the motor barge *Beverley Brook* and the lighter *Helmsdale* were also driven ashore, as was the barge yacht *Diligent*. When it came to clearing up the mess, the *Viper* was salvaged by the strange expedient of drilling holes in the hull to let the water out at low tide.

The *Greater London* had five calls in one day again on November 23rd 1938, in a south-westerly gale. So strong was the wind that day that boys on roller skates holding pieces of canvas above their heads as sails were keeping pace with cars along Southend front. The larger Leigh cockle boats were forced to shelter off Shoeburyness and a dozen small craft riding at moorings between Leigh and Southend were swamped. Shortly after 9 a.m. the lifeboat was called to the Southend-owned barge *Neepawa*, which was making for the Corporation loading jetty with a cargo of ballast. The two men on board were taken off just before the barge went down about half a mile east of the pier.

About noon the lifeboat was summoned to the assistance of the 20 ft. skiff *Audrey* near the Chapman light, and one man was taken off. In the meantime the lighter *Helmsdale* was in trouble off the gasworks jetty, against which the *Beverley Brook* was being battered by the waves. The lighter sank about a quarter-mile from the shore and her skipper took to the boat, which he tethered to the lighter's cabin. He had to bail furiously for an hour until the lifeboat got back and picked him up. Later the *Greater London* was called to the schooner *Westward*, which was dragging her anchor, but the lifeboat's help was not needed.

Shortly before 6 p.m. the lifeboat was called out a fifth time when a barge was reported to be in trouble off Warden Point, on the Isle of Sheppey, but nothing was found.

A couple of days later the *Greater London* was in service again when a westerly gale caused smaller craft to run for shelter and bigger vessels to anchor and ride out the squalls. Boatbuilder Frank Peters saw torch signals from the barge *Glenmore*, at anchor a quarter of a mile off the Camper Road jetty; she had a cargo of ballast for Shoeburyness. The lifeboat took the skipper and the mate from the *Glenmore*, then the lifeboatmen turned their attention to the barge *Lord Roberts*, which was sinking fast in the short seas being kicked up by the gale. On the way back after taking two men off this barge the lifeboat took the crew from yet a third barge, the *T.F.C.*, which was lying between the Corporation jetty and the pier.

When the tide dropped, all three barges were left on the mud and next day workmen were busy draining the water out of them, while a P.L.A. salvage craft was at work on the *Neepawa*, which had gone down in deeper water.

For the service to these three barges, very skilfully carried out in the worst of weather, and in exceptionally difficult and severe conditions, Coxswain Page was awarded the second service clasp to his Bronze Medal. He was only the second man in the past 30 years to win two medals in one year.

Came 1939 and the lifeboats went to war, as did Southend pier itself which was taken over by the Navy on August 25th, a few days before the actual declaration of war. It held a strategic position in the Thames estuary, in the approaches to the country's greatest port, and it was suddenly no longer just a landing place for carefree holidaymakers who had come down by paddle steamer from the Metropolis. Where the day trippers had spent freely of their cash at the pierhead, naval personnel busied themselves in ripping out the paraphernalia of peacetime and in setting up the headquarters of the Thames and Medway Control. Southend became the assembly point for East Coast convoys, and masters of ships joining the convoys came to the pierhead for briefings; between 1939 and the end of hostilities in 1945 no fewer than 3,367 convoys, consisting of 84,297 ships, were to be assembled there.

A defence boom intended to prevent enemy surface craft or submarines gaining access to the estuary was built across from Shoeburyness to the Kent shore. This was hung on piles driven into the sands where the water was shallow and from lighters moored in the deep-water channels. It proved an additional hazard and a barrier to friendly vessels as well as to enemy craft, and on occasions the lifeboat had difficulty in negotiating the 'gateways' through which ships and boats had to make their way.

For the duration of the war naval control was an important link in the chain of which Southend lifeboat formed a part. It was not naval control but Margate coastguards, however, who alerted the lifeboat when the Dutch steamer *Spaarndam* was mined three miles east-nor'-east of the Tongue lightvessel on November 27th 1939. The *Greater London*, together with the Margate lifeboat *Lord Southborough*, went to the spot but the crew had been taken off by a Trinity House pilot cutter and the *Greater London* found only one of the ship's boats, which she took in tow.

The lifeboat had several fruitless calls at this period. On December 9th rockets were reported from the vicinity of the Mouse lightvessel, but nothing was found, and on January 20th 1940 clouds of smoke were seen out at sea but again the lifeboat could find nothing. On this occasion the cold was severe and the lifeboat returned covered with ice from stem to stern.

Like other lifeboats from the East Coast, the *Greater London* took part in the evacuation from Dunkirk in May and June, 1940. Manned by a naval crew, she was put to work in Dunkirk harbour and in the last hours of the evacuation the *Greater London* saved the surveying vessel H.M.S. *Kellett*, which was trying to embark 200 French soldiers from the mole. It was 1.30 a.m. on June 4th, and the harbour was being cleared of what remained before blockships were sunk at the entrance. If she were not to be trapped, the warship must leave at once, but was caught by some underwater obstruction. Her bows were touching the beach, but when she was put astern one of her screws caught the obstruction and she could not be moved. It was at this juncture that the *Greater London* passed by, full of soldiers. She was hailed from the bridge of H.M.S. *Kellett*. Could she give them a pluck off the beach? She did, thus saving the vessel and her crew from capture.

That was by no means her only war service. On October 19th the *Greater London* was called out when F. T. Everard's motor coaster *Aridity* struck a mine near the East Oaze lightvessel. The crew had already been taken off but the lifeboat stood by until the *Aridity* sank in shallow water. Little more than a week later the lifeboatmen had another experience of wartime destruction when the collier s.s. *Letchworth* sank as a result of enemy action in the same area. Of the 18 men picked up by a naval vessel, five were seriously injured, a dozen had minor wounds and one was dead. The lifeboat landed the survivors.

The lifeboat crew had a spate of very hard but rewarding work in November and December 1940. Coxswain Page received the Institution's thanks on vellum and his crew received framed letters of thanks for their work on November 12th and 13th, when between 9 a.m. one day and 3.30 a.m. the next they went out seven times to the aid of vessels in distress in a south-westerly gale. The lifeboat was at sea almost continuously for 18 hours in very rough conditions, and the work was made harder by the fact that there were only the coxswain and six men instead of the full crew.

The first call came at 9.15 in the morning when the steel billy-boy *Halcyon*, by then converted to a fully-powered motor vessel, was seen driving before the gale with one anchor down a quarter of a mile west of the pier. It was impossible in the conditions then prevailing to get her to a safer anchorage, so two lifeboatmen got aboard her and helped the crew of four to let go her second anchor. This stopped her drifting.

The lifeboat got back to her station at 11.15 a.m., but a quarter of an hour later she was called out to a number of barges which were in a very dangerous position. The crews refused to leave as they thought they could ride out the gale, but asked the lifeboat to keep an eye on them. It was a quarter past midday when the lifeboat arrived back at the pierhead, and at half past the boat put out for the third time, to the barge *Tintara*, which was signalling for help a mile south-west of the pier. She was submerged in terrific seas and completely out of control. Coxswain Page very skilfully manoeuvred the lifeboat alongside and took off the barge's two-man crew.

Just 25 minutes after getting back and landing them, the *Greater London* was out again on what proved to be the most difficult and dangerous service of the day. One of the 200-ton lighters from which the defence boom was hung had been torn away and was driving down the fairway, which was crowded with shipping held up by the gale. There was only one man on the lighter, and there was grave danger of what the coxswain called 'a big mix-up' with the anchored shipping. Coxswain Page, handling the lifeboat with great skill, got alongside the drifting lighter and took off the man. It seemed impossible that, in those heavy seas, the lifeboat could take the much larger vessel in tow, but the coxswain realised that the effort must be made and, after long and valiant efforts, this was done.

Within half an hour of getting back to their station the crew were taking the *Greater London* out for a fifth time, to the Trinity House tender *Argus*, which had struck a mine and sunk. Her mast could be seen above the water about three-quarters of a mile east of the Mouse lightvessel with men clinging to the rigging. It was now dark, but with the help of their searchlight, the lifeboatmen found the mast of the sunken steamer. There was, however, no sign of the crew, who must all have been washed off; it was impossible for anyone to have lived in that heavy sea, so the lifeboat returned to its station.

Again there was only a half-hour before the next call came, and at 9.45 p.m. the lifeboat put out for the sixth time. The gale was still blowing fiercely and there seemed a danger that some of the barges at anchor might be carried through the pier. The lifeboat succeeded in taking in tow the *Tintara* and the Horlock 'ironpot' *Port light*, which were unmanned, and took them to a safe anchorage. She arrived back at 3.30 a.m. and the exhausted crew had five hours to get a little rest before the next call came. It was again to the *Halcyon*, which they had succeeded in getting safely anchored the day before, and this was taken in tow to another anchorage. It was over 26 hours since she first put out when the *Greater London* got back to the pierhead.

The Southend lifeboatmen had an exceptionally long trip beginning on November 16th, when a message came that a ship was sinking close by the Mouse lightvessel. It was the steamer *Dagenham*, which had been mined and was then lying in shallow water close to the East Cant buoy. The lifeboat took aboard 15 of her crew, leaving the captain and six officers aboard, and at the request of the commodore of the convoy of which the *Dagenham* formed a part returned to stand by the sinking ship after landing the rescued men. When she got back there were tugs standing by, but they left at 6.30 p.m. The lifeboat stood by all through the night, keeping watch on the vessel, which had by then sunk.

At 4.30 in the morning the lifeboat went alongside and took off the six officers, but the captain would not leave, so a lifeboatman was left to keep him company while the *Greater London* landed the six officers at Southend. Later in the morning the lifeboat was called to the Canadian steamer *Sherbrooke*, but when the *Greater London* reached her tugs had her in tow and she was not in need of assistance, so the lifeboat returned to the *Dagenham*. The captain was still determined to stay on board as salvage tugs had arrived, so the lifeboat re-embarked its crew member and returned to its station after 28 hours. Three days later the superintendent of the owners asked if the lifeboat would to out to the *Dagenham* again as no other boat was available, so the *Greater London* put the ship's chief engineer on board the *Dagenham* to join the captain, who was still in the sunken vessel's wheelhouse. Eventually the *Dagenham* was refloated and towed away to be repaired.

It was not very long before the lifeboatmen had another busy spell, for on December 5th a westerly gale was blowing when information came that a barge was burning flares about two miles off Warden Point. The *Greater London* went out and found the barge *Nelson* anchored in the four-fathom channel on the edge of the Middle sand, in a very dangerous position. She was leaking and the two men on board had passed a terrible night. Coxswain Page put some of his crew on board the barge to lift the anchor and then the lifeboat towed her to Southend.

The gale was still blowing next morning when the lifeboat was called out to the help of another barge. That launch was followed by seven others to the help of six other barges and a steamer's crew. The lifeboat was out in the gale almost continuously for nearly 36 hours.

First, the lifeboatmen found the spritty *Glencoe* in distress half a mile west of the West Shoebury buoy, but her skipper refused to leave and as the lifeboat was unable to take the barge in tow in such a very heavy sea the *Greater London* returned to her station to await any further calls. They were not long in coming, for a quarter of an hour later she put out to the barge *Mistley*, which was a quarter of a mile west of the

West Shoebury buoy and driving towards the sands. By the time the lifeboat reached her she had sunk and her crew of two were clinging to the mast, with heavy seas breaking over them and threatening at any moment to wash them away.

A quarter of an hour after landing the *Mistley*'s crew at the pierhead the *Greater London* put out to Goldsmith's spritties *Decima* and *Cambria*, which had driven on to the sand to the westward of the pier and sunk. There were two men clinging to the rigging of each barge. Coxswain Page could see that it was going to be no easy task to pluck them from their perches, for only the length of a barge separated the two sunken vessels, which had their anchors down and their cables stretching to windward, out to sea. He decided to anchor the lifeboat and to drop down stern first to the barge which was nearer the shore.

This took the lifeboat herself very close to the shore, but she got alongside the barge and took off the two men. Then she moved seawards again, picking up her cable as she went, to go to the second barge. While she was doing this the cable got to the wrong side of her, passed under her as she moved forward, and fouled one of the propellers, but Coxswain Page managed to go alongside the second barge and take her crew from the rigging by using only one engine.

Then the cable fouled the second propeller. The lifeboat swung stern on to the seas, which broke right into her after cockpit. She was helpless. Her coxswain signalled for a tug, but in that sea no tug could come to his help. The bowman at once offered to go over the side and attempt to clear the propellers, but the coxswain knew he would have been drowned and would not allow him to attempt it. Instead he set some of his men to the oars and the others to haul on the cable.

It was now that the depleted crew most felt the need of the other two men—there was a crew of only six instead of the normal eight—but by rowing hard they were able to take the strain off the cable and, by tremendous hauling, to get some of it on board. The lifeboatmen then made fast to it a strong rope from the bows and cut the cable where it was attached to the propellers. The lifeboat swung head to wind, and the immediate danger was over.

But the lifeboat had still to get away. She carried no sails, unlike the earlier motor lifeboats which had retained sails as a secondary means of propulsion. She had only a light mast intended simply to carry a steaming light. But Coxswain Page remembered that under the cable they had coiled over the after canopy as a protection against shell and bomb splinters was a piece of canvas to prevent the rope from rubbing the varnish. It was only 14 ft. by 7 ft., but it would get her home, the coxswain decided. So extra stays were put on the mast and the make-shift sail was set. The flood tide on the lifeboat's lee bow helped to keep her from being blown back on to the shore and the coxswain successfully sailed the *Greater London* to the pier.

Other vessels were in distress and the lifeboat was needed at once. Before she could go out, though, her propellers had to be cleared. The lifeboat was hauled round the pier to her slipway. Getting her on to this in such weather proved no easy task, and eight times she was washed off before the crew, displaying fine seamanship, could haul her out and free the rope.

By now the crew were wet through and very tired, but there was no rest for them and within half an hour the lifeboat had been launched for the fourth time. This was to the *Glencoe*, whose skipper had refused to leave her when the lifeboat went out to his aid in the morning, and the *Nelson*, which the lifeboat had towed in the day before. The *Glencoe* was two miles south-east of the pier. Her crew were at the pumps, but again the master refused to leave and asked for a tow. Again the cox-swain had to refuse, for the *Nelson* was waiting for his help.

He found her crew standing in the rigging, wet and cold. He rescued them and then returned to the *Glencoe*, whose determined skipper again asked for a tow. Lifeboatmen helped him to get his anchor and the tow started. An attempt was made to set some of the barge's sails, but they were soon blown to ribbons.

Then the mate of the barge was washed overboard. Luckily he caught hold of a chain and was hauled aboard by the lifeboatmen, but he was injured. When the lifeboat and her tow came inshore Coxswain Page ordered the barge's anchors to be let go and he went aboard her, telling the captain that he would have to take him off. The skipper replied that the injured mate might go, but he would stay, and he went down into the cabin, followed by the coxswain. They found a foot of water over the floor. Then at last the skipper consented to leave, and he and his mate boarded the lifeboat. Half an hour later the *Glencoe* sank.

It was now 8.30 in the evening, and the lifeboat had been out almost continuously for nearly eleven hours. Any crew might be forgiven for saying they had had enough, but half an hour later the *Greater London* put out for the fifth time.

The *Cambria*, from which the lifeboat had taken the crew early in the afternoon, had been carried against the pier. She was full of water, but her cargo of timber kept her afloat. The lifeboatmen did their best to prevent the barge from breaking through the pier, but they found it impossible to move her and had to return to the lifeboat station to refill the boat's petrol tanks. At four o'clock next morning they tried again, but again they could not move the waterlogged barge, so they went ashore, borrowed a pump from the fire brigade and returned to pump out the barge. That done, they were able to tow the *Cambria* round the pier. The lifeboat did not get back to her station until 11.30 that morning, December 7th.

Even then she had not finished, for a quarter of an hour later she was launched to the aid of a wrecked steamer, the *Houston City*. Her crew had already been taken off by another steamer, however, and the lifeboat arrived back at 3.45 in the afternoon, only to put out again a quarter of an hour later to yet another barge, the *Verona*. Her sails had blown away and her topmast and sprit had broken. The gale was now moderating, and the lifeboat towed in the barge with her two men on board.

It was nine in the evening before she was back in her boathouse. She had then been out almost continuously for nearly 36 hours, yet she had still not finished with barges. Next day she went out to bring in the *Glencoe*, which she succeeded in doing after her crew had pumped the barge out with their borrowed fire pump and had thrown overboard much of her deck cargo. On the 9th the *Greater London* towed in the *Cambria* and on the 11th, after again using the fire pump, her crew salved the *Decima*.

For the services to the *Cambria*, *Decima*, *Glencoe* and *Nelson* on December 6th, Coxswain Page was awarded the Silver Medal and the five members of his crew the Bronze Medal.

In 1941 German efforts to strangle Britain by preventing essential supplies reaching the country by sea were, in spite of the counter-measures taken by the Royal Navy, causing a serious loss of tonnage. Vessels which had escaped the attention of U-boats were all too often sunk near land by magnetic or acoustic mines laid by German aircraft at night, although the work of converted trawlers and specially built motor minesweepers in keeping open swept channels proved invaluable in keeping losses to a minimum. Others were bombed and strafed by marauding aircraft.

Most of the minings were outside the boom, the sweepers normally being successful in keeping the inner channels clear of mines. But at six in the morning of July 13th 1941, the coxswain received a message that a ship had been mined about half a mile from the pier. The *Greater London* went out and stood by the s.s. *Collingdoc* while she was towed to shallow water, then took off 17 of her crew of 24 and landed them before returning to take off the captain and second officer; others remained on the beached steamer.

At the end of 1941 or the beginning of 1942 the *Greater London* was replaced by the reserve lifeboat *J. B. Proudfoot*, which served at Southend until the middle of 1945 when the *Greater London* returned. The *J. B. Proudfoot* had been built at Cowes in 1924 as the *H. F. Bailey* for the Cromer station, but was replaced there by a second *H. F. Bailey* after only a few years' service. While the *J. B. Proudfoot* was at Southend the *Greater London* was doing duty at Yarmouth, Isle of Wight.

Launching the lifeboat presented difficulties when naval control asked the coxswain to assemble his crew on October 26th 1942, for two wreck-lifting lighters which were bringing in a wreck had moored across the end of the slipway only sixty feet away from it and had allowed the wreck to take the bottom there. As soon as the lifeboat left the slipway her engine had to be put full astern to check her way and prevent her from running into the moored lighters.

Somewhere in the east anchorage a vessel was sending out a radio S.O.S., and to the east of the pier the firefloat *Norwich Belle*, in peacetime a pleasure steamer taking holidaymakers for trips in Yarmouth Roads and out to the Scroby sand, and a tug were ashore. Coxswain Page decided to take the *J. B. Proudfoot* first to the vessel sending out the S.O.S. as he felt the other two would be in no immediate danger in spite of the southerly gale and rough sea.

The night was dark, with a mist over the sea, but after searching along the edge of the sands for only a few minutes the lifeboatmen found the converted yacht barge *Thoma II*, a handsome clipper-bowed, counter-sterned spritty built as a yacht by that builder of thoroughbred barges, Howard, at Maldon in 1909. Commandeered early in the war, she was at this period in the service of the Port of London Authority.

Lying broadside to the seas, which were breaking around her, the *Thoma II* was rolling heavily. The tide was low and there was not enough water for the lifeboat to get under her lee, so Coxswain Page anchored to seaward and dropped down towards her on his cable. When she was about thirty feet from the barge the lifeboat grounded on a sandbank and, broadside to the seas, began to roll heavily with the seas breaking over her all the time.

In spite of this the lifeboatmen threw a heaving cane to the barge, rigged the breeches buoy, and hauled the crew of the yacht through the surf to the lifeboat. There were eight of them, and it took half an hour to get them all into the lifeboat.

There was nothing to do then but wait for the tide to flood and refloat the lifeboat. Once she came off the sand the lifeboat returned to the pier to land the eight rescued men before setting off to the assistance of the tug and the firefloat. When she got to them it was found that another tug was helping the grounded tug, which had lost her propeller, so the coxswain took the lifeboat to the help of the firefloat.

The *Norwich Belle* was lying on the sand in about a fathom of water. The coxswain ran out a spare anchor with a rope cable attached to it, passed the cable to the firefloat and then put three of his crew on board the firefloat. They put the cable round a small hand capstan and, with the firefloat's engine working, hauled her off the sands.

Considering that both the launch and the rescue of the crew of the barge yacht were skilfully carried out in face of considerable difficulty the Institution awarded Coxswain Page its thanks on vellum and made additional awards to the coxswain and crew.

Three months afterwards the lifeboat not only saved another of the firefloats but saved Southend pier from being cut in two, a matter of no little importance to the naval authorities whose headquarters would have been cut off if the pier had been breached. The first call came at 11 a.m. on January 31st 1943, when naval control reported that the firefloat *Gladys* was driving on to the sands west of the pier. Coxswain Page was in London in connection with a broadcast the B.B.C. was producing on the work of a lifeboat station in wartime, so the second coxswain took command of the *J. B. Proudfoot* when she was launched into a whole gale. The *Gladys* was being swept by heavy seas as she hit the sand and the crew of 14 were all very seasick, but the lifeboatmen got them on board the lifeboat and landed them safely.

Next morning the coxswain was told by the piermaster that the *Gladys* had been washed against the pier and was crashing into it. As he made his way to the boathouse, Coxswain Page was met by a naval officer who asked that something should be done to prevent the firefloat from cutting the pier in two. Taking a second anchor on board the lifeboat, he anchored the *J. B. Proudfoot* to windward of the float and fired a line to her. The line was seized by a lifeboatman who courageously dropped from the pier on to the firefloat, which had already ripped a number of girders from the pier; the torn girders lay on her deck. With the help of another vessel the lifeboat towed the *Gladys* clear and took her under the lee of the pier, where naval control took her over.

When it was not such craft as the firefloats there were naval craft of all kinds getting into trouble. In August 1943, it was one of the landing craft which were gathering for the coming invasion of Europe which had to be towed in by the lifeboat, and in December the *J. B. Proudfoot* was again out to landing craft. Then in May 1944, the lifeboat took two men from the naval boat *Basil George* which went ashore in squally conditions half a mile east of the pier, and in October the *J. B. Proudfoot* performed a most useful if unspectacular service in taking a convoy commodore and boarding officer from the naval tug *Leigh* when she went ashore on the Nore sand in foggy weather and putting the officers aboard their ships.

H.M.S. *Leigh* was again in trouble on December 10th that year, when the *J. B. Proudfoot* found her on a sandbank 150 yards east of the pier and driving towards it. The tug had only a few fathoms of chain and the anchor and windlass were practically useless. The crew of six and three boarding officers were ready to leave, but two lifeboatmen jumped on board the tug in hopes of getting her away from the pier and to a safe anchorage. The lifeboat ran out an anchor which held the tug and saved her from striking the pier, then took off the three boarding officers, a task which took some time because of the awkward conditions and the darkness. After landing them she returned to the tug and stood by until the *Leigh* was able to move under her own power.

Just a week after this service the lifeboat crew was called together early in the morning to take the *J. B. Proudfoot* to the Maplin sands where naval control reported that H.M.S. *Jeanie Deans*, well known to thousands of people in peacetime as a Clyde paddle steamer, had been driven ashore on the Blacktail spit in a strong southerly gale.

The *Jeanie Deans* was being swept by heavy seas, and her officers asked the lifeboat to stand by so that she could pass a towrope if a tug arrived. At daylight H.M. Tug *Advice* came on the scene, but by now the tide had turned and the sea had increased. In spite of this the lifeboat anchored, veered down to within 120 feet of the *Jeanie Deans* and, with a line-throwing gun, a line was thrown on to the paddle steamer's port bow.

Seeing what was happening the tug let go her anchor and dropped down towards the lifeboat, and eventually the lifeboatmen managed to transfer a cable to the tug. A tow-wire was bent to the cable and the crew of the *Jeanie Deans* pulled it on board, but after the tug had been pulling for some time the wire broke. The *Jeanie Deans* was by now in a bad way, continually pounded by heavy seas.

To attempt to get another tow-wire on board the lifeboat had to go head first to the sands, the seas making a clean sweep over her, filling the canopy and carrying two of her crew under it. Fortunately they were not injured. In spite of all the difficulties, the tow-wire was replaced and the *Jeanie Deans*' own towing wire was also taken to the tug. By taking the weight on the two wires the tug succeeded in hauling the paddle steamer off the sands just before high water.

Without the assistance of the lifeboat the *Jeanie Deans* and her complement of 90 could not have been saved, for the tug could never have got near her because of her draught and the heavy seas.

The lifeboat was asked to stand by until the *Jeanie Deans* reached sheltered waters, but while escorting her the lifeboat crew saw two barges flying distress signals about half a mile west of the South Shoebury buoy. As the *Jeanie Deans* was now able to proceed under her own steam the lifeboat went to the barges, the *Ida*, of Ipswich, and the 'ironpot' *Xylonite*, of Harwich. They were anchored in a very exposed position, were badly damaged and unmanageable. The crews had been exposed to the gale for hours and were wet, cold and hungry, and it was only with difficulty that the lifeboatmen got them from their vessels.

After landing the four men at Southend pier, where they were given hot food and dry clothing, the lifeboat put out again to save the barges and their valuable cargoes of flour.

Sometimes the services rendered by the lifeboat were of a more mundane kind. The day after Boxing Day 1944, naval control asked for food to be taken out to the drifter *Ocean Lover*, one of many East Anglian herring boats taken over by the Admiralty and employed on various duties, which had been lying a mile east of the boom for four days owing to thick fog. The *J. B. Proudfoot* went first to the vessel guarding the boom, to discover that this vessel was also short of food.

Those were busy days at Southend. On January 2nd 1945, the lifeboat was called out when the wartime standard ship *Fort St. Paul* and H.M.S. *Ben-my-Chree*, in peacetime plying between Liverpool and the Isle of Man, were in collision in Sea Reach in thick fog.

Later the same month the *J. B. Proudfoot* was called to the m.v. *Eileena*, a small coaster which had parted from her anchors during a nor' wester and was being battered against the spiked floats and chains of the defence boom about a mile south of the Medway gate. Coxswain Page found it impossible to get the lifeboat alongside because of the gale and a strong ebb tide, but it was clear that something had to be done, and done quickly, if the *Eileena* was not to become a total wreck. After several attempts the lifeboat got a line to her, a towrope was made fast and the lifeboat towed the *Eileena* clear of the boom, whereupon the coaster's crew were able to start her engine.

The coxswain reported to naval control after escorting the *Eileena* to Southend and was at once asked to go out again to four tugs, the *TID 111*, *TID 112*, *TID 113* and *TID 115*, small wartime-built steam tugs employed mainly on dock and river towage duties. The previous night they had gone to the south shore for shelter from a southerly gale, but the wind had got round to the north-west and was knocking up a very rough sea. The lifeboatmen found the tugs close to the sands between the Yantlet and Jenkin buoys, their crews short of food and their bunkers low. The tugmasters asked the lifeboatmen to send a radio message to Southend, as a result of which the tugs were directed to go to Sheerness.

The problems posed in lifeboat work are ever-changing; no two rescues are alike, and each service calls for a different plan of action. Coxswain Page and his crew can have had little idea of the peculiar difficulties they would be faced with when they were called from their beds a little after four on the morning of February 3rd 1945. A strong southerly wind was blowing the sleet into the faces of the lifeboatmen as the *Greater London*, back at last from her wartime banishment to the Isle of Wight, was launched to the assistance of H.M. Drifter *Silver Seas*, which had been abandoned and was adrift just to the west of the pier. As in so many similar cases there was a danger of the vessel driving into the pier and possibly cutting through it, but in such weather it was impossible for the lifeboat to take the drifter in tow.

Instead the coxswain decided the only answer was to anchor the vessel until such time as towing was possible, so he let go the lifeboat's main anchor and, paying out 70 fathoms of cable, steamed near enough to the *Silver Seas* for three of his crew to jump aboard her. Then, using a heaving line, he transferred the lifeboat's cable to the drifter, the three men making it fast and so preventing the drifter from driving on to the pier.

About 5.10 a.m. the steam tug *TID 7* came on the scene. The lifeboat's towrope was made fast to the drifter and the lifeboat made towards the tug. The lifeboatmen were trying to pass the rope to the tug when that vessel came astern and crashed into the lifeboat, but despite this and after a good deal of effort the towrope was transferred to the tug. Then it got round the tug's propeller.

She was too close to the pier to anchor, so her crew hauled her alongside the drifter and made her fast. Going alongside the drifter, the lifeboatmen passed over the lifeboat's spare anchor and cable and this held the two vessels until they grounded as the tide fell.

At low water Coxswain Page went out to the tug and the rope was cleared from the propeller and the anchor was carried further out.

That afternoon the lifeboat again went out to the two vessels. Three lifeboatmen were put on the drifter and another boarded the tug, and then, under the direction of Coxswain Page, the tug made fast her towrope. The lifeboat's towrope was in turn passed to the tug and after much effort the drifter was refloated.

With the end of the war, work began on restoring the pier to its peacetime appearance and on removing the defence boom across the estuary. This in itself was no simple task and on one occasion an accident to the men engaged on the demolition of the boom led to the lifeboat being called out. It was a message from the garrison at Shoebury that alerted the lifeboat crew on April 5th 1946, when seven men were stranded on the boom a mile and a half out to sea and an amphibious vehicle trying to rescue them was smashed against the boom. The lifeboatmen found the men clinging to the boom, exhausted. They had great difficulty in getting the lifeboat close to the boom owing to the strong tide, but in the end they succeeded in taking the men aboard, one at a time, over the bows.

The following month the sailing barge *Astrild* crashed into the boom defence in the darkness and got into serious difficulties through her cargo shifting. The lifeboat crew found her in danger of sinking, but three lifeboatmen got on board and worked her clear of the boom, and the lifeboat then towed her to a safe anchorage.

Later in 1946 the lifeboat carried out a search for the body of Geoffrey de Havilland, chief test pilot of the aircraft firm bearing his name, after his D.H. 108 tailless aircraft had blown up in the air and crashed into the Thames Estuary off All Hallows, opposite Canvey Island. The lifeboatmen found oil patches in the position where the fuselage of the aircraft was later located, but nothing could be found of the pilot.

The *Greater London* played a leading part in attempts to save the crew of the submarine *Truculent*, which sank near the Nore flats on

January 12th 1950 after a collision with a small Swedish tanker which was outward bound from London for Ipswich. The *Truculent*, a T-class patrol submarine built at Barrow under the war programme in 1942, was proceeding on the surface north-west of the Red Sands tower when she was rammed on the starboard side in the torpedo room by the tanker, the *Divina*. She was returning to Sheerness after trials following a refit, and besides her crew of six officers and 55 ratings there were 18 dockyard workmen on board.

The forward part of the submarine was flooded and she sank rapidly. As she went down four officers and a rating were swept into the water; those who were below—apart from those trapped by the inrush of water after the collision—concentrated in the engine room and the after compartment. Just over half an hour after the collision the Davis escape apparatus was brought into use. But nothing was known ashore of the collision, for the *Divina* was not fitted with long-range radio and could not contact the shore with her radio telephone.

The first signal to be received was from the Holland-Amerika liner *Almdijk* about three-quarters of an hour after the collision, but this only reported that she was rescuing men from the water—the five men washed off the bridge as the submarine went down—and it was not realised on board the *Almdijk* that a submarine had been sunk. Even those on board the *Divina* believed that their ship had collided with some small craft and did not realise until later that a submarine had been sunk with a large number of men on board.

Not long after eight everything was ready and the first man left the submarine. Several left one after the other, some wearing Davis escape apparatus and others without it—there had been no time to grab the sets from the part of the submarine which had been flooded as she sank. The men tried to keep together after surfacing, but the racing ebb tide prevented that.

Then at 8.15 p.m. the *Almdijk*'s second signal gave the naval authorities their first intimation of the disaster. 'H.M. Submarine *Truculent* sunk N.W. Red Sand tower between X4 buoy and E. Piles buoy . . . all ships please keep lookout.'

But by this time all survivors must have left the submarine. The delay of over an hour in beginning rescue operations was the cause of tragedy. The submariners had surfaced fairly soon after the accident because they heard the sound of propellers above them and because with so many crowded into the two compartments it was feared that the air would quickly become bad. Unseen in the darkness the swimmers who had escaped were borne away by the swift-running ebb—some bodies were later picked up in the Barrow Deep by the destroyer H.M.S. *Zest*. Forty-eight of the crew and all but two of the dockyard men were lost.

The *Greater London* was the first lifeboat to respond to the call and was the second vessel to arrive on the scene after the *Almdijk*'s S.O.S. Coxswain Page reported to the commanding officer of the Hunt-class escort destroyer H.M.S. *Cowdray*, which had arrived a little earlier. After finding out what he could Coxswain Page returned to the lifeboat and they set off to search for survivors and for the other vessel known to have been involved.

While searching the lifeboatmen saw a tanker of about 600 tons approaching and went alongside to make inquiries. What Coxswain Page saw when he got on board the tanker was ten survivors from the *Truculent*, and he realised in a flash that this was the vessel involved in the collision.

Coxswain Page was seriously handicapped because the Swedish skipper was the only man who spoke any English, and he had the greatest difficulty in making himself understood. The survivors were transferred to the lifeboat as soon as possible and the lifeboat made for H.M.S. *Cowdray*.

After putting the survivors on board the naval vessel the lifeboat took a naval officer to the *Divina*. As Coxswain Page and the naval officer climbed to the tanker's deck they saw a number of bodies lying on the deck, still wearing their Davis submarine escape apparatus. 'We immediately started artificial respiration, but by then it was too late,' the coxswain reported.

By the time the lifeboat got back to the *Cowdray*, after the coxswain had got as much information from the tanker's skipper and the pilot as he could, a large number of vessels, including other warships, had arrived on the scene. A plan was drawn up for a comprehensive search. The two lifeboats, from Southend and Margate, were ordered to the flanks of the line of ships and instructed to search the buoys for survivors who might have climbed on to them.

The sweep was soon under way and the whole estuary was a blaze of light as searchlights swept across the dark water, but no further survivors were seen. By this time the actual wreck area had been pinpointed and the vessels were again mustered, arrangements being made for a depth charge to be blown as a signal for any survivors left in the submarine's hull to use their escape apparatus and come to the surface.

Before the explosion Southend lifeboat stationed itself practically over the sunken submarine, but nobody surfaced nearby.

Before resuming the now almost hopeless search the lifeboatmen went on board the *Cowdray* for a meal in the petty officers' mess. It was only after 24 hours work, during which time the lifeboat used more than a hundred gallons of fuel, that the *Greater London* was recalled.

The lifeboatmen received special thanks from the Navy for their help, and before returning to the pier the Commander-in-Chief, Nore, Admiral Sir Henry Ruthven Moore, received Coxswain Page on board the *Cowdray* to thank him for his efforts.

The *Greater London* was called out several times during the great gale of January 31st-February 1st 1953, which caused the loss of the Irish Sea ferry *Princess Victoria*, the Lowestoft trawler *Guava*, and other vessels, and caused a tidal surge which swept down the East Coast swamping coastal towns and villages and inundating large areas of low-lying land.

At 9.23 a.m. on the 31st she put out to the motor barge *Adriatic* which was in trouble a couple of miles from the pier, and in the afternoon she was launched twice, the first time to pick up the 40 ft. Cornish ketch *Wanderlust*, drifting seawards a quarter of a mile north-west of the pier, and the second time to recover the bawley *Patience*, rolling heavily and adrift off Chalkwell, three-quarters of a mile west of the pier.

Then at 10.18 p.m. the 19,000-ton Norwegian tanker *Kosmos V* signalled that she was aground on the Shoebury sands, two and a half miles from the pier. In view of the combined menace of a high wind with gale force gusts and a prematurely high tide Coxswain Page decided to launch the *Greater London* for the fourth time that day, and the lifeboat stood by the *Kosmos V* until she refloated and anchored safely in deeper water.

While the lifeboat was standing by the Norwegian tanker the tide had been rising inexorably above the predicted level. At 12.45 a.m. the *Greater London* went out a fifth time, to see how things were at Canvey Island, which is entirely below sea level and protected by a sea wall and dykes built largely by Dutch engineers. While on the way up the estuary Walton coastguard sent a radio message to the lifeboat about the plight of two men marooned on the roof of the Leigh sewerage works and the lifeboat changed course for Two Tree Island, in the direction of Leigh. As she changed course the creek walls on Canvey

Island were collapsing, letting through a wall of water which caught many people before they could make for the upper floors of their houses and even wrecked some of the houses.

There was little the lifeboatmen could do, and the *Greater London* returned to her station to stand by for further duty. Not much later she was out again, taking the boarding boat with her, to Canvey Island but with the tide now ebbing she was able only to make contact with people who were in no danger.

Plans were made that day for the lifeboat to reconnoitre Foulness Island on the night tide, as amphibious vehicles had failed to get through to the island earlier. One DUKW had been abandoned and another had lost itself on Havengore Island—all landmarks on the relatively featureless landscape were covered with floodwater. Shortly after 9 p.m. on February 1st the *Greater London* left Southend, taking two small boats with her, making her way round Shoeburyness, past Wakering Stairs and Havengore Head towards Fishermen's Head, some 15 miles from her station. It was difficult to find a suitable place to land the small boats, and the lifeboat went right into the mouth of the Crouch before such a place could be found. Here the lifeboatmen discovered a great breach in the sea wall on the north side of the island, and soon the two small boats had been taken through the gap on to the island. The lifeboatmen rowed over to a farm, but they found the people there unwilling to leave until daylight.

The tug *Mediator* and the minesweeper *M.M.S. 1728* left the Medway on the morning of February 2nd with orders to stand by off Foulness to receive people evacuated by the lifeboat, but in fact the lifeboat and other rescue craft took the evacuated residents to Burnham. Altogether the lifeboat evacuated 25 people.

The *Greater London* (*Civil Service No. 3*) was withdrawn from Southend in 1955 and was replaced by a Watson cabin type lifeboat, the *Greater London* (*Civil Service No. 30*). The £32,163 she cost was a gift to the Institution from the Civil Service Lifeboat Fund. After a period in reserve the old boat went to Ecuador in 1957 to help found that country's lifeboat service.

The passing of sail and the conversion of the remaining spritties to power and the building of new motor barges did not mean that the lifeboat was any less needed. In 1955 Coxswain Page was awarded the Institution's thanks on vellum for the rescue of the crew of the motor barge *Fence*, which had grounded on the West Barrow sand while on her way from Fingringhoe on the Colne with a cargo of sand for London River.

'They reached us about midnight and had three goes taking us off', said Skipper George Petersen. 'It was very difficult. Never in my life have I seen such cool men as that lifeboat crew. They might have been on a mere white-weeding trip by the way they behaved.'

The lifeboatmen tried to take the barge in tow, but a five-inch rope parted and they had to abandon her—though only for the time being. Next day the lifeboat went out again, found her seven miles from where they had left her anchored and towed her to Southend.

That was Coxswain Page's last notable service. He retired in December 1955, and then became honorary boathouse attendant—Southend is one of the few stations to have an attendant at the lifeboat-house. He died in 1962 at the age of 71.

Peter Gilson, who became coxswain in 1956, and his crew won the collective thanks of the Institution on vellum for an unusual service to a Dutch coaster which was on fire in the Thames Estuary. It was an hour before low water on December 8th 1962, when the coastguard told honorary secretary Percy Garon that the 198-ton coaster *Temar* was at anchor near the No. 2 Sea Reach buoy and was on fire. The *Greater London* was launched to her aid, but the *Temar* reported that the tanker *Mobil Enterprise*, two tugs and a local boat had gone to her assistance and she did not need the help of a lifeboat. In spite of this Coxswain Gilson decided to make for the Dutchman as he believed lives might be in danger. Subsequent happenings showed that his decision was the right one.

As the lifeboat circled the *Temar* the crew could see a man's head jammed in a cabin porthole on the *Temar*'s starboard quarter. A ship's boat from the *Mobil Enterprise* had already put aboard the tanker's chief engineer, her second officer and a deckhand with an asbestos suit and oxygen breathing apparatus, but they could not reach the man's head either from the coaster's deck or from the ship's boat.

The wind had increased to a near gale, and one of the tugs had a line aboard the *Temar*. Coxswain Gilson asked the tugmaster to hold her in such a position that she afforded a lee and then secured the lifeboat under her starboard quarter. All hands on board the *Temar* were then mustered on deck to man the head and stern warps.

Two of the lifeboatmen, Bowman A. Martin and F. Emery, boarded the *Temar* and took turns at keeping the man's head out of the porthole by pulling on his ears, while Coxswain Gilson and Motor Mechanic

John Polkinghorn administered oxygen each time the lifeboat came up on a wave. They were hindered by smoke and could only give the oxygen for a few seconds at a time as the lifeboat rose and fell in the swell.

It was then decided to cut away the *Temar*'s deck to allow the man, who was by now in a very distressed condition, to be pulled out. This was done with equipment on one of the tugs. As the deck was cut above the cabin a jet of water was played on to the man to protect him from the effects of the molten metal falling on his back.

When the deck had been cut away the chief engineer of the *Mobil Enterprise*, dressed in an asbestos suit, pulled the man out from the cabin. He was taken aboard the lifeboat and landed on Southend pier, where a doctor and an ambulance were waiting.

The lifeboat then put out again to the *Mobil Enterprise*, from which she picked up two men, two women and two children, all survivors from the *Temar*. She put them ashore at Sheerness before returning to Southend.

That is not the end of the story of the Southend lifeboat by any means. She is still as busy as any lifeboat on the British coast, for scarcely a month goes by when she is not called out on some service or other, whether it be taking a doctor out to attend a sick seaman, landing an injured man, or rescuing the crew of a fishing boat or yacht in trouble in rough weather. In 1956 the lifeboat was joined by an inshore rescue boat based in a hut on a jetty just to the east of the pier, from which she is launched by a rather complicated hoist.

Up to the end of 1966 the Southend lifeboats had rescued a total of 636 people. It seems that this total is certain to increase in the years to come.

BOAT RECORD, SOUTHEND

Years on station	Length, Breadth, Oars/crew	Type, Weight Cost	Year built (Off. no.) Builder	Boat's name, Donor Authority
1879-1889	25′ 7′ 3″ 8 db/	SR £352	1879	*Boys of England and Edwin J. Brett* "Boys of England" magazine R.N.L.I.
1889-1891	25′ 7′ 3″ 8 db/	SR	1882 (259) Woolfe, Shadwell	*Boys of England and Edwin J. Brett*[1] As previous boat R.N.L.I.

1885- 1899	34′ 7′ 6″ 10/13	SR £300	1885 (33) Woolfe, Shadwell	*Theodore and Herbert*[2] Mrs. Frances Sophia Smith, Lisheen, Co. Cork R.N.L.I.
1899- 1923	38′ 12′ 12/	N & S 6t 11 £865	1899 (430) Thames Iron Works, Blackwall	*James Stevens No. 9* Leg. James Stevens, R.N.L.I. [Birmingham
1923- 1924	41′ 11′ 12/15	W 9t 14	1903 (505) Thames Iron Works, Blackwall	*William Roberts*[3] Leg. William Roberts, R.N.L.I. [Manchester
1924- 1927	43′ 12′ 6″ 10/13	W 11t 13 £1,694	1904 (535) Thames Iron Works, Blackwall	*Charlie Medland*[4] Leg. Charles Brandon [Medland, London R.N.L.I.
1928-1941 1945-1955	48′ 6″ 13′ 6/8	Ramsgate motor 21t 3 £8,668	1928 (704) White, Cowes	*Greater London* (*Civil Service No. 3*) Civil Service LB Fund R.N.L.I.
1941- 1945	45′ 12′ 6″ -/6	W cabin motor 17t 9	1924 (694) White, Cowes	*J. B. Proudfoot*[5] (ex-*H. F. Bailey*) R.N.L.I.
1955-	46′ 9″ 12′ 9″ -/8	W cabin motor 21t 7 £32,163	1955 (921) Groves and Guttridge, Cowes	*Greater London* (*Civil Service No. 30*) Civil Service LB Fund R.N.L.I.

1 Ex Weston-super-Mare; to Southport as boarding boat, 1891.
2 Transferred to Workington, 1899.
3 Reserve boat, originally at Littlehaven.
4 Transferred from the Mumbles.
5 Reserve boat, originally at Cromer.

SERVICE RECORD, SOUTHEND

Boys of England and Edwin J. Brett Lifeboat

1880	Mar.	2	Smack, gave help	
1881	Oct.	14	Barge *Butcher*, of Rochester, saved	2
			Barge *Maria*, of Rochester, saved	2
			Barge *Violet*, of Rochester, gave help	
1882	Oct.	24	Brig *August*, of Griefswald, assisted to save vessel and	7
1886	Dec.	8	Schooner *Wietska*, saved	5
1887	Sept.	2	Brig *Avenir*, of Kragero, assisted to save vessel and	7

SOUTHEND No. 2

Theodore and Herbert Lifeboat

1887	Oct.	10	Schooner *Venus*, of Teignmouth, stood by	
	Dec.	17	Barge *Sultan*, of Rochester, stood by	
1893	Dec.	8	Barge *J. D. Gough*, of Rochester, assisted to save vessel and	2
1894	Jan.	19	Schooner *Vectic*, of Teignmouth, saved	4
1895	Nov.	11	Barquentine *Durango*, of Hamburg, assisted to save vessel and	10
1896	Oct.	6	Men in danger on pier extension works, saved	7
1897	Mar.	3	Barge *Ada*, of London, gave help	
			Barge *Charles Little*, of Rochester, saved	2

1898	Mar.	24	Schooner *Hugh Miller*, of Inverness, saved	6
1899	Jan.	12	Barge *Whitwell*, of London, assisted to save vessel and	3
	Feb.	12	Schooner *Billow*, of Inverness, landed 4 from Mouse lightship	

James Stevens No. 9 Lifeboat

1900	May	14	Schooner *Sydney Smith*, of Caernarvon, saved vessel and	7
	Dec.	20	Barque *Delta*, of Fredrikstad, saved vessel and	10
1901	Mar.	25	Barque *Grethe*, of Kragero, stood by	
1903	Mar.	2	Barge *Edith*, of Rochester, saved	2
			Barge *Mary Ann*, of London, saved	2
1904	Dec.	30	Barge *Sandy*, of London, saved vessel and	3
			Barge *Kimberley*, of Ipswich, saved vessel and	2
1906	Nov.	30	Barquentine *Claremont*, of Dublin, assisted to save vessel and	7
1908	Jan.	8	Barge *Dorothea*, of Harwich, saved vessel and	2
	Dec.	10	Ketch *Canterbury Belle*, of Plymouth, saved vessel and	4
	Dec.	28	Barque *Kappa*, of Brevik, assisted to save vessel and	10
1909	Feb.	5	Schooner *Prosperity*, of London, saved	5
	Dec.	3	Barge *Redan*, of Faversham, saved	1
1910	Oct.	2	s.s. *Duke of Abercorn*, of Southend, gave help	
1911	Jan.	30	Schooner *Cimbri*, of Gloucester, assisted to save vessel	
	Apr.	6	Schooner *Martin Gust*, of Riga, saved vessel and	7
	Sept.	30	Ketch *Invicta*, of Harwich, saved vessel and	3
	Sept.	30- Oct. 1	Barge *Audacious*, of Colchester, saved vessel and	2
1913	Jan.	11	Boat of s.s. *Jeanie Hope*, of London, gave help	
	Feb.	8	Barge *The Sisters*, of Ipswich, landed 2 from Mouse lightship	
	Mar.	8	Boat of ship *Aravia*, of Nystad, saved	2
	Apr.	4	Schooner *Antje*, of Hamburg, saved	1
	Apr.	20	Barge *William and Richard*, of London, saved	2
	June	10	s.s. *Palma*, of Belfast, saved a boat and	10
	Dec.	27	s.s. *Mona*, of Newcastle, stood by	
			s.s. *Ophir*, of Glasgow, saved a boat and	6
1914	Dec.	29	Tug *Woodcock*, of London, assisted to save vessel and	4
			Barge *Eliza*, of Rochester, saved vessel and landed 2	
			Barge *John Tinnoth*, of Rochester, assisted to save vessel and	2
1915	Jan.	1	Barge *Outsider*, of Rochester, stood by	
1916	Jan.	13-14	Admiralty hopper, stood by	
	Mar.	27-28	Barge *Victoria*, of Ipswich, gave help	
	June	4	Schooner *Francis and Jane*, of Harwich, gave help	
1920	Jan.	11	Barge *Irex*, of Rochester, landed 2	
	Dec.	4	s.s. *Graf Waldersee*, of London, gave help	
1921	Nov.	6	Barge *Avon*, of London, saved	2
1922	Mar.	8	s.s. *Orcades*, of London, assisted to save vessel	
	Dec.	13	Schooner *Elizabeth Bennett*, of Swansea, saved	5
1923	Apr.	21	Barge *Jeffie*, of Rochester, saved	2

Charlie Medland Lifeboat

1925	Jan.	31	Barge *Albert*, of Rochester, saved	2
	Apr.	16	Cutter *Vision*, of Brixham, saved	3
			Barge *Hand in hand*, of Rochester, saved	2

Greater London (*Civil Service No. 3*) Lifeboat

1928	Aug.	20	Launches *Sussex Queen* and *Sussex Maid*, saved boats and	4
	Sept.	12	Yacht *Pamela*, of Southampton, stood by	
	Oct.	6	s.s. *City of Lancaster*, of Liverpool, stood by	
	Nov.	16	s.s. *Hunterfield*, of Leith, stood by	
1929	July	5	Yacht *Otter*, of Westcliff, assisted to save vessel	
	Nov.	11	Barge *Onward*, of Whitstable, saved	2
1930	Aug.	2	Houseboat *Diadem*, of Faversham, saved boat, landed 5	
	Nov.	23	Barge *Marion*, of London, landed 2	
			Barge *Emily*, of Ipswich, landed 2	
			Barge *Charles and Esther*, of Wakering, saved	2

1931	Mar.	1	Boat of steam tanker *Pass of Ballater*, of London, landed 10	
	May	3	Small boat *Edna*, of Leigh, landed 2	
	July	25	Yacht *Dorothy*, of Southampton, landed 5	
1932	Jan.	6	Barges *Cader Idris* and *Stanley Margetts*, of Rochester, saved	3
			Barge *Lady Mary*, of London, stood by	
	Jan.	15	Barge *Conqueror*, of Rochester, saved	2
	July	13	Motor boat *ex-Bacchus*, of Burnham, gave help	
	Oct.	29	Barge *New Hope*, of Maldon, saved	2
1934	Mar.	4	Barge *Claxfield*, of Queenborough, saved	2
	Apr.	20	s.s. *Penton*, of Newcastle, gave help	
	Aug.	21	Barge *Florence Myall*, of Rochester, gave help	
	Oct.	14	Barge *Scarborough*, gave help	
	Oct.	21	Canoe, saved canoe and	1
			Motor boat of Queenborough, landed 3	
1935	Feb.	16	Barge *Sunrise*, gave help	
	May	19	Yacht *Squib II*, of Westcliff, saved yacht	
1936	Jan.	21	s.s. *Ubari*, of Tallinn, stood by	
	May	22	Pinnace from schooner yacht *Westward*, of London, gave help	
	Aug.	19	Yacht *Mary*, of Leigh, gave help	
	Oct.	25	Yacht *Cachalot*, of Burnham, gave help	
			Yacht *Spray*, of Portsmouth, gave help	

J. B. Proudfoot Reserve Lifeboat

1937	Mar.	15	Barge *Norseman*, gave help	
	May	18	Motor yacht *Evadne*, of Glasgow, gave help and landed	5
	June	12	s.s. *Royal Archer*, of Leith, stood by and landed 49	

Greater London (*Civil Service No. 3*) Lifeboat

1937	Nov.	10	Motor boat *Coronia*, of Shoebury, saved boat and	3
	Dec.	13	Barge *Gannet*, of Gillingham, saved	2
1938	Feb.	15	Barge *Pacific*, of London, saved	2
			s.s. *Dunnett Head*, of Leith, landed 1	
	Feb.	16	Barge *Pacific*, of London, saved vessel	
			Boat of s.s. *Corton*, of London, saved boat and 5	
	Apr.	29	Small fishing boat of Shoeburyness, landed 3	
	June	1	Yacht *Wimpie*, of Southampton, saved	2
			Barge *Glenrosa*, of Harwich, saved	3
			Barge *Audrey*, of London, saved	2
	Oct.	2	Two fishing boats of Shoeburyness, landed 17	
	Nov.	19	Motor yacht *Dawn*, of Southend, landed 3	
	Nov.	23	Barge *Neepawa*, of Rochester, saved	2
			Fishing boat *Audrey*, of Leigh, saved	1
			Lighter *Helmsdale*, of London, saved	1
	Nov.	26	Barge *T.F.C.*, of Rochester, saved	2
			Barge *Glenmore*, of Rochester, saved	2
			Barge *Lord Roberts*, saved	2
1939	Mar.	20	Barge *British Oak*, of Rochester, saved barge and	2
	Mar.	21	Yacht *Don Pat*, of Sheerness, saved yacht and	2

The Brothers Reserve Lifeboat

1939	June	21	Yacht *Redshank*, of Shoeburyness, saved yacht	

Greater London (*Civil Service No. 3*) Lifeboat

1939	Oct.	23	s.s. *Copeland*, of Glasgow, stood by	
	Nov.	26	Barge *Five Sisters*, stood by	
	Nov.	27	s.s. *Spaarndam*, of Rotterdam, saved ship's boat	
1940	May-June		Took part in Dunkirk evacuation with a naval crew	
	Oct.	19	m.v. *Aridity*, of London, stood by	
	Nov.	1	s.s. *Letchworth*, of Newcastle, landed 18	
	Nov.	12	m.v. *Halcyon*, of Hull, saved vessel and	4

1940	Nov.	12	Barge *Tintara*, of London, saved	2
			Admiralty boom defence lighter, saved lighter and	1
			Barges *Tintara* and *Portlight*, of London, saved barges	
	Nov.	13	m.v. *Halcyon*, of Hull, gave help	
	Nov.	16-17	s.s. *Dagenham*, of London, saved	21
	Nov.	20	s.s. *Dagenham*, of London, Put officer aboard	
	Dec.	5	Barge *Nelson*, of London, saved barge and	2
	Dec.	6	Barge *Mistley*, of Ipswich, saved	2
			Barges *Cambria* and *Decima*, of London, saved	4
			Barges *Glencoe* and *Nelson*, of London, saved	4
			Barge *Cambria*, of London, saved barge	
	Dec.	7	Barge *Verona*, of London, saved barge and	2
	Dec.	8	Barge *Glencoe*, of London, saved barge	
	Dec.	9	Barge *Cambria*, of London, gave help	
	Dec.	11	Barge *Decima*, of London, saved barge	
	Dec.	27	m.v. *Araby*, of Liverpool, saved	28

J. B. Proudfoot Reserve Lifeboat

1941	Jan.	15	Barge *Ferret*, of London, stood by	

Greater London (*Civil Service No. 3*) Lifeboat

1941	Apr.	10	Barge *Rowland*, of London, saved barge	
	May	19	s.s. *Winkfield*, of Glasgow, landed 54	
	July	13	s.s. *Collingdoc*, of Newcastle, saved	19
	Oct.	18	Ex-Belgian defence vessel *Andre Marcel*, saved vessel and	7

J. B. Proudfoot Reserve Lifeboat

1942	Apr.	7	N.F.S. firefloat, saved vessel and	7
	Apr.	27	Barge *Emma*, of Rochester, saved vessel and	1
	Oct.	26	P.L.A. vessel *Thoma II*, of London, saved	8
			N.F.S. firefloat *Norwich Belle*, saved vessel and	7
1943	Jan.	31	N.F.S. firefloat *Gladys*, of London, saved	14
	Feb.	1	N.F.S. firefloat *Gladys*, of London, saved vessel	
	Aug.	30	*L.C.P. 1058*, gave help	
1944	May	5	Naval boat *Basil George*, saved	2
	Oct.	8	H.M.S. *Leigh*, gave help	
	Dec.	10	H.M.S. *Leigh*, saved vessel and landed 3	
	Dec.	17	H.M.S. *Jeanie Deans*, assisted to save vessel and	90
			Barges *Ida*, of Ipswich and *Xylonite*, of Harwich, saved barges and	4
	Dec.	27	Drifter *Ocean Lover*, took out food	
			American s.s. *Sun Yat Sen*, put officer aboard	
			Canadian s.s. *Calgary*, put officer aboard	
1945	Jan.	2	American s.s. *Fort St. Paul*, and	
			Admiralty vessel *Ben-my-Chree*, stood by	
	Jan.	19	m.v. *Eileena*, of London, saved vessel and	3
			Four tugs, gave help	

Greater London (*Civil Service No. 3*) Lifeboat

1945	Feb.	3	H.M.S. *Silver Seas*, and a tug, assisted to save vessel	
	Oct.	28	s.s. *Moorwood*, of London, gave help	
1946	Feb.	20	Barge *Bassildon*, of London, saved barge	
	Apr.	5	Seven men marooned on boom defence, saved	7
	May	2	Barge *Astrild*, of London, saved barge and	2
	Aug.	28	Yachts *Modwena*, of Birmingham, and *Sandpiper*, of Christchurch, saved yachts and	4
	Sept.	14	Yacht *Alexandra*, of Whitstable, saved	2
	Dec.	17	Lighters *Van Tromp* and *Halo*, landed 2	
1947	Apr.	4	Motor yacht *Arthur Rogers*, escorted yacht	
	Apr.	8	Barge *Thetis*, of London, saved barge	

Milburn Reserve Lifeboat

1947	Apr.	18	Motor yacht *Prince Siki*, saved yacht	
	Apr.	23	Motor barge *Maid of Munster*, of Rochester, saved	2
			Barge *Asphodel*, of London, saved	2
			Barge *Adriatic*, of London, saved	2
	June	8	Yacht *Joass*, of Chatham, gave help	

Greater London (*Civil Service No. 3*) Lifeboat

1947	Sept.	16	Yacht *Bellbo*, of Upnor, landed 2	
1948	Jan.	7	Barge *Thelma*, of Rochester, saved	2
			Barge *Thelma*, of Rochester, saved barge	
	Apr.	15	m.v. *Kingfisher*, of London, landed a sick man	
	July	4	Yacht *Diver*, of Gillingham, gave help	
	Oct.	17	Two ex-landing craft, saved boats and	3
	Nov.	17	m.v. *Quo Vadis*, of Groningen, landed 10	
	Dec.	9	Barge *G.C.B.*, of Rochester, saved	2
			Barge *G.C.B.*, of Rochester, saved barge	
1949	Mar.	1	Barge *Asphodel*, of London, saved	2
			Barge *Kitty*, of Harwich, saved	2
			Barge *Kitty*, of Harwich, saved barge	
	May	18	Yacht *Drifter*, of Leigh, saved yacht	
	June	7	Yacht *Cormorant*, of Westcliff, saved yacht	
	Aug.	1	Yacht *Dorie*, of London, saved yacht and	3
	Nov.	15	m.v. *Mercurius*, of Amsterdam, gave help	
	Dec.	11	Lighter *Mediator*, of London, saved vessel	
1950	Jan.	12	H.M. Submarine *Truculent*, gave help	
	Feb.	12	Barge *Brian Boru*, of London, saved	2
	Feb.	26	People marooned on Dead Man's Island, River Medway, saved	3
	May	5	Boat, gave help	
	May	27	Yacht *Pinta*, of Gillingham, saved boat and	2
	Nov.	20	Motor barge *Ada Mary*, of Rochester, saved	2
	Nov.	26	s.s. *Selnes*, of Oslo, stood by	
	Nov.	29	m.v. *Cub H*, of Hull, gave help	
	Dec.	1	Motor tanker *Guidesman*, of London, gave help	
1951	Sept.	1	Motor yacht *Joles II*, of London, saved	2
	Sept.	30	Motor yacht *Snafelle*, of London, escorted yacht	
	Oct.	3	s.s. *Yewcroft*, of Glasgow, stood by	
1952	Mar.	29	m.v. *Britannia*, of Southend, saved vessel	
	May	11	Two boats, saved a boat and	5
	June	1	Yacht *Passat*, of London, saved yacht and	3
	Aug.	9	Motor yacht *Santa Lucia*, of Brussels, saved yacht and	2
	Aug.	19	*Hopper No. 19*, of London, saved two boats and	8
	Oct.	5	Yacht *Wyvern*, of Shoeburyness, landed 2	
	Nov.	7	Barge *Savoy*, of Dover, stood by	
	Dec.	17	Motor barge *Nellie*, of Faversham, saved	2
1953	Jan.	31	Ketch *Wanderlust*, saved boat and	1
			Fishing boat of London, saved boat	
			Tanker *Kosmos V*, of Sandefjord, stood by	
	Feb.	1-2	People marooned by floods on Foulness Island. Gave help and landed 25	
	Mar.	1	m.v. *Baalbeck*, of Oslo and Great Nore forts, gave help	
	Mar.	4	m.v. *Northumbrian Coast*, of Middlesbrough, saved vessel	
	Mar.	7	m.v. *Prittlewell*, of London, gave help	
	July	29	s.s. *San Cirilo*, of London, put doctor on board	
			Yacht *Sutta*, stood by	
	Aug.	2	s.s. *Beaconsfield*, of London, put doctor on board	
	Aug.	3	s.s. *Beaconsfield*, of London, landed a sick man	
	Aug.	29	Yacht *Escape*, of Upnor, stood by	
	Sept.	10	s.s. *Wandle*, of London, put doctor on board	

1953	Sept.	21	Yacht *Tinker*, of Upnor, saved yacht and	3
			Motor yacht *Bryter*, of Benfleet, escorted yacht	
	Oct.	8	R.A.F. launch, escorted vessel	
	Dec.	24	m.v. *Alf Everard*, of London, landed 7 from another vessel	
1954	Mar.	7	s.s. *Cape Clear*, of Glasgow, gave help	
	Mar.	9	m.v. *Kenrix*, of Hull, took out doctor	
	Mar.	14	Salvage vessel *Juniper*, of London, saved vessel and	4

Lord Southborough (*Civil Service No. 1*) Reserve Lifeboat

1954	May	15	Punt *Brumas*, saved boat and	3
	June	10	Motor cruiser *Ubique*, of Erith, saved dinghy and	3
	June	10-11	Motor cruiser *Ubique*, of Erith, saved cruiser	
	June	18	Speedboat *Blue Peter*, of Margate, saved boat and	2

Greater London (*Civil Service No. 3*) Lifeboat

1954	Sept.	20	Yacht *Scamp*, of Gravesend, saved yacht and	3
	Nov.	18	s.s. *Caprella*, of Panama, took out doctor	
	Nov.	27	s.s. *Olivia*, of Mariehamn, took out doctor	
	Dec.	22	Lighter *Gerrard*, of London, gave help	
	Dec.	23	Fishing vessel *Rolling Wave*, transferred injured man to motor vessel	
1955	Mar.	4	s.s. *Mitcham*, of London, took out doctor	

Greater London (*Civil Service No. 30*) Lifeboat

1955	May	18	Aux. yacht *Stormalong*, of London, saved yacht and	3
	June	12	Aux. yacht *Blotto*, of Felixstowe, saved yacht, landed 3	
	Oct.	21	m.b. *Fence*, of London, saved	3
	Oct.	22	m.b. *Fence*, of London, saved vessel	
1956	Mar.	3	Tanker *Tank Queen*, of Oslo, landed a sick man, **saving his life**	1
	July	28	Raft, saved raft	
	July	29	Yacht *Ocean Breaker*, saved boat and	6
	Sept.	1	Motor barge *Maria*, of Rochester, saved boat and	2
	Sept.	28	m.v. *Shawford*, gave help	
	Oct.	29	Barge *Colonia*, of Colchester, saved	2
	Dec.	14	m.v. *Raycrest*, of London, saved vessel and	3
1957	Apr.	19	Yacht *Ava*, of London, gave help	
	June	6	Yacht *African Queen*, saved boat and	4
	Aug.	10	Motor cruiser *Flamingo*, of Leigh, saved boat and	3
	Aug.	11	Yacht *Sandling*, of London, saved boat and	5
	Aug.	18	Motor cruiser *Jochelle*, saved boat and	4
	Oct.	20	Yacht *Vandra*, of London, saved boat and	3
	Dec.	17	s.s. *King David*, of London, landed injured man	1
1958	Feb.	26	Tug *Hawkstone*, of London, gave help	
	Apr.	23	m.v. *Militence*, of Rochester, landed injured man	1
	June	27	Motor boat from s.s. *Gladstone Star*, saved boat and	9
	July	13	Yacht *Leila*, of Grays, saved boat and	3
	July	14	Yacht *Hilary Anne*, of Leith, gave help	
	July	18	Sailing dinghy, saved boat	
	July	28	Motor cruiser *Marjon II*, saved boat and	6
	Aug.	1	Yacht *Bridget*, of Felixstowe, saved boat and	2
	Sept.	7	Fishing boat *John Patrick*, gave help	
	Dec.	28	Motor tanker *Lonn*, of Bergen, landed sick man	
1959	Feb.	20	s.s. *City of Perth*, of London, landed sick man	
	Apr.	3	s.s. *Ramso*, escorted vessel	
	Apr.	26	Motor barge *Glenwood*, of London, saved vessel and	2
	May	23	Two yachts, saved yachts and	4
	July	7	Yacht *Annasona*, of Chatham, stood by	
	July	8	Landed two bathers	
	July	23	s.s. *John Orwell Philips*, took out doctor, landed a body	

1959	Aug.	18	s.s. *Persic*, of Southampton, took out doctor	
	Dec.	24	Tug *Exchange*, of Southampton, saved boat and	2
1960	Jan.	19	Tug *Exchange*, of Southampton, gave help	
	May	4	Dinghy, landed 2	
	June	12	Yacht *Dolphin*, of Southend, gave help	
1961	Feb.	18	Boat from Swedish vessel *Sira*, gave help	
	Aug.	8	Barge *Knox*, of Rochester, gave help	
			Yacht *Cee Jay*, of London, gave help	
	Sept.	4	Motor cruiser *Honolulu*, saved boat and rescued a cat	4
	Sept.	26	Motor tanker *Aureity*, of London, landed a sick man	
1962	Jan.	3	m.v. *Devonbrook*, of London, landed a sick man	
	Jan.	13	m.v. *Tjoba*, of Groningen, escorted vessel	
	Apr.	6	Reported cries for help, landed 1	
	Aug.	26	Three yachts, saved	9
	Sept.	11	Motor cruiser *Istar*, and ship's boat, saved two boats and landed 2	
	Oct.	11	Motor barge *Adieu*, of Ipswich, saved barge and	2
	Dec.	8	m.v. *Temar*, of Mepple, landed 6, rescued	1
1964	June	7	Yacht, saved boat, rescued	3
	Sept.	28	Radio Invicta on Red Sand Fort, landed injured man	
	Oct.	1	Motor cruiser *Baltic Lily*, saved boat	
	Nov.	7	Radio Invicta, landed sick man	
	Nov.	10	Ferry boat *Edith*, of London, gave help	
	Nov.	11	Russian m.v. *Vetlugales*, landed a sick man	
	Dec.	21	Radio Invicta, took out stores	
1965	Feb.	21	Motor cruiser *Cedefore*, gave help	
	Mar.	18	Dredger *Foremost*, landed injured man	
	Apr.	21	Motor bawley *Gladys*, gave help	
	Apr.	24	Dinghy, saved boat	
	May	8	Motor cruiser *Toby*, saved boat and	3
	July	4	Fishing vessel *Alice and Ella*, gave help	
	July	11	Yacht *Zoe*, saved boat and	2
	Aug.	26	Fishing cruiser *Gay Buccaneer*, gave help	
	Sept.	1	Ketch *Isabella*, of Fowey, gave help	
	Sept.	8	Motor cruiser *Vulfrano*, saved boat and	2
	Sept.	26	Motor cruiser, saved boat and	4
	Sept.	28	m.v. *Sir Joseph Rawlinson*, landed body	
	Nov.	1	m.v. *Clarity*, of London, gave help	
	Nov.	27	Motor barge *Rodent*, of Rochester, gave help	
	Dec.	29	Ketch *Scarab*, saved	3
1966	Mar.	8	Tanker *Springwater*, landed injured man, thereby saving his life	1
	Apr.	7	Motor cruiser *Pandora*, gave help	
	July	16	Fishing vessel *British Angler*, gave help	
	July	24	Sloop *Skim*, saved boat and	3
	Aug.	31	Motor cruisers *Apex* and *Gloria J*, gave help	
	Dec.	11	Motor cruiser *Janorma*, saved boat and	3

THE LIFEBOAT BUILDERS

FROM the earliest days Essex has played an important part in the development of the lifeboat. According to tradition, Lionel Lukin—who may reasonably claim to be one of the inventors of the lifeboat—made early experiments with models of his 'unimmergible boat' on the Doctor's Pond at Dunmow, his birthplace. The patent which he obtained in 1785 shows that he anticipated the design of later lifeboats in many ways: the specification mentions 'projecting gunnel, in a faint curve, towards the water, so as not to interrupt the oars in rowing' and 'inclosures, or bulkheads, made watertight, or filled with cork, or other light material that will repel the water' as well as a false keel of cast-iron.

In 1786 Lukin fitted up a coble which was used for some years at Bamburgh, Northumberland, for lifesaving work. He thus contributed to the first lifesaving station on the British coast.

The sailing lifeboat which he designed for the Suffolk Humane Society in 1807 on the lines of the beach yawls used at Lowestoft for salvage work and for attending on shipping in the roads followed this specification. The forerunner of the successful Norfolk and Suffolk type of lifeboat, this vessel had wide cork wales and casks lashed under the thwarts and along the insides of the gunwales to provide buoyancy.

Another Essex man who influenced the development of the lifeboat was George Palmer, of Nazeing Park, whose design was adopted by the National Life-boat Institution soon after its formation. Palmer went to sea at the age of 14, making his first voyage in the East Indiaman *Carnatic* in 1786, and it was an accident to a boat of which he was in charge in 1788 that turned his attention to the design of boats, particularly with reference to their stability and the means of keeping them afloat when swamped.

He made his last voyage in 1799 and three years later entered into partnership as an East India merchant in London. He became connected with the Lifeboat Institution in 1826, two years after its formation, and two years later published a pamphlet on 'A new plan for fitting all boats so that they may be secure as Life Boats as the shortest notice'. He was deputy chairman of the Institution for upwards of a quarter of a century, resigning in 1853 shortly before his death at the age of 81.

His boat was double-ended, with upright stem and stern, with a cork fender 4 in. wide and 4 in. deep along the sides. She had detached air cases of wood along the sides and in bow and sternsheets, but no means of freeing herself of water, the design depending on a small internal capacity for holding water.

According to the model entered in the Northumberland prize competition in 1851, she was clench built of elm and fir, copper fastened. The model was of a boat 25 ft. long, 6 ft. 9 in. broad and 3 ft. 3 in. deep, pulling five oars single-banked. She drew only 15 in. with 20 men on board and cost only £75 at that time. Not all the boats built to Palmer's design were exactly alike, but the dimensions seem to have differed only slightly, all being 26-28 ft. long.

This type gave way in 1852 to a new self-righting boat designed for the Institution by James Peake, assistant master shipwright of Woolwich Dockyard. This boat combined all the best points of the models submitted for the Northumberland premium, was double-ended and diagonally built of two thicknesses of rock elm, copper fastened. Many of this type and of the Institution's later self-righting type were built at Limehouse by Forrestt's, a company which had been established as far back as 1788.

There had been lifeboats built at Limehouse before T. & W. Forrestt, as the firm was known at the time, began building for the National Life-boat Institution. Harton had built five lifeboats between 1828 and 1835, and Taylor had built two in 1837 and 1839, but it was not until 1852 that Limehouse became the centre of lifeboat production.

It was not only lifeboats for the Institution that were built by Forrestt's, for in 1860 they were said to have recently built eight lifeboats, of Peake's design but of a smaller size than the Institution boats, to be carried on four Spanish warships being built by Samuda Brothers at Millwall.

A year or so earlier Norway Yard had been busy on lifeboats for the Norfolk coast following the Institution's takeover from the Norfolk Shipwreck Association. About a dozen lifeboats under construction were destroyed or damaged by a fire at the yard in the summer of 1858, but by 1860 the firm had sent new boats of Peake's design to Cromer, Mundesley, Bacton, Palling, Winterton and Yarmouth to replace the old ones on those stations.

Among the large number of craft built by Forrestt's for the R.N.L.I. were the *Honourable Artillery Company* for Walton and the second *Albert Edward* for Clacton, the first R.N.L.I. lifeboat to be fitted with a drop keel.

In 1888 Forrestt's took over the shipyard at Wivenhoe, on the Colne, earlier owned by J. Harvey & Co. and J. E. Wilkins. They continued to build boats—and to lose money—at Norway Yard until 1897. Some of the lifeboats built between 1888 and 1896, when they gave up building for the Institution, were constructed at Wivenhoe Shipyard, but they are now chiefly remembered for the shallow-draught craft built for service in the colonies and various overseas countries. Many of these special-service craft, designed to be taken apart for shipment and later re-erected by unskilled labour on the bank of some African or South American lake or river, were built to the order of the Crown Agents for the Colonies, and this connection seems to have brought the firm some of its lifeboat work.

In 1896 Forrestt's received an order to build a large self-righting lifeboat for the Government of New South Wales. This, shipped to Australia in 1897, was probably the copper-sheathed lifeboat, Yard No. 345, which the records show to have cost £1,221 15s. to build, including £50 11s. 9d. for delivery to Goole, whence she was apparently shipped. This boat was 50 ft. long, with a beam of 12 ft. 6 in., unusually large for a lifeboat in those days.

In this case the price of £1,606 left a fair profit, but the Wivenhoe yard's 'black book' reveals that lifeboat work seldom showed a profit. The Institution demanded a high standard of workmanship and materials, and it seems that at this period there was keen competition for such orders, with the result that prices quoted by the successful tenderers were on the low side. Vessel No. 150, a 39 ft. boat for the R.N.L.I., is recorded as costing £628 7s. 11d. to build 'plus 15 per cent—£722 13s. 2d.'—presumably an allowance for overheads and profits—but the price of £597 meant a loss of more than £30 on the cost of materials and labour alone. The firm did very little better on the next order, a 34 ft. R.N.L.I. lifeboat, for the price of £442 left the very narrowest margin over the actual cost of £441 19s. 9d. But if one took account of the 15 per cent, which brought the cost to £508 5s. 8d., the transaction showed quite a heavy loss. Vessel No. 152, another 34 ft. boat, cost £496, almost £70 more than the price the R.N.L.I. paid for her.

It is thus not surprising to find it recorded that Forrestt & Son Ltd.

traded at a loss between 1892 and 1898, and at only a small profit for the next three years, after which they found themselves in financial difficulties.

The firm did not deal only with wooden boats, for the Wivenhoe yard also turned out a 31 ft. steel self-righting lifeboat, Yard No. 192, which cost £235 to build.

In 1909 the Turkish government ordered two self-righting pulling and sailing lifeboats from the Wivenhoe shipyard. They were to be exactly like the R.N.L.I. self-righters, then being built by the Thames Iron Works, but the plans of these boats could not be obtained. So a shipwright and a draughtsman from Wivenhoe travelled to Hunstanton, on the Norfolk coast, to take off the lines of the R.N.L.B. *Licensed Victualler*, a 35 ft. boat built by Thames Iron Works in 1900. So satisfactory did these first two boats prove that the yard later received a further order from Turkey for two more. The hull of one was complete when war broke out in 1914; work on them stopped and the completed hull was eventually made use of as a shed.

In 1912 G. Rennie & Co. moved to Wivenhoe from Greenwich because of labour troubles on the Thames, and an amalgamation resulted in the two concerns becoming the Rennie Forrestt Shipbuilding, Engineering and Dry Dock Co. Ltd. About this time the yard turned out a steam lifeboat, the *Moseley*, for the Crown Agents for the Colonies. Designed by Sir E. J. Reed, this was for service at Lagos and was said at the time to be the fastest lifeboat afloat.

Successive changes of name followed the war period, the firm becoming first the Rennie, Ritchie and Newport Shipbuilding Co. and then Otto Anderson and Co. (London) Ltd. The yard was closed, like so many others of similar size, during the depression of the 1930's.

A Thames-side yard which played a notable part in the development of British lifeboats as well as an even more notable part in the history of shipbuilding was that of the Thames Iron Works at Canning Town, extending some way up the Essex side of Bow Creek. The first iron shipbuilding on the Thames was carried on at this yard by Ditchburn and Mare in the 1840s, and it was here that H.M.S. *Warrior*, the first sea-going ironclad in the world, was built between 1859 and 1861. In rather more than half a century the yard turned out about a thousand vessels.

Towards the end of 1895 the Thames Iron Works found it impossible to obtain satisfactory wooden boats from other suppliers, so it was decided to open a boatbuilding department. This new branch began by building 21 boats for the Japanese battleships *Fuji* and *Shikishima*, and then went on to construct three 56 ft. vedette boats ordered by the Admiralty.

In April 1896, the firm entered into an agreement with the R.N.L.I. to build all their boats under the Institution's superintendence and inspection at a percentage over and above the cost of labour and material. The first built of these were a 34 ft. boat for Rhyl and a 35 ft. boat for Port St. Mary, completed successfully in the contract time of three months, followed the next year by a 37-footer for Newhaven and a 35-footer for Sutton. Four more followed in 1898 and six more in 1899, numbers increasing markedly in succeeding years.

Problems were numerous in the new department at the outset. In the early months of 1898 an order was received for a 40 ft. lifeboat for the Cape of Good Hope, but the boat shop was sadly in need of work. Then in June the foreman boatbuilder left, to be replaced by one of Forrestt's men.

By the end of 1898 the boatbuilding ship was "very full", but the increased activity brought its own problems. The yard had to keep a large stock of timber on the premises to ensure that there was always an adequate supply of sound, well-seasoned timber for the construction of the R.N.L.I. craft. But in June 1899, when there were no fewer than 16 lifeboats building for the Institution, it was proving difficult to get the necessary skilled hands, while there was also a shortage of oak crooks for the stems.

In the opening years of the present century there were seldom fewer than a dozen lifeboats in hand. In June 1901, there were only nine although as many had been delivered in the previous three months; shortly afterwards it was reported that five more had been delivered and that 12 were building. It is of interest to note that one of the boats delivered in 1901, the *Constance Melanie*, a 35-footer which was sent to Coverack, had been provided from the legacy of Mr. F. E. Hills, the eldest brother of the company's chairman.

Early in 1902 there were 19 lifeboats under construction and though the following year the lifeboat work was said to be 'fairly satisfactory' but 'not so busy as usual' there were still 16 in hand. In the 16 years Thames I.W. was engaged on lifeboat work more than 200 lifeboats joined the R.N.L.I. fleet from the Canning Town yard.

The company was an interesting one and was in the forefront of social as well as technical development. After a period of industrial unrest in 1889-91 a system of profit-sharing was introduced—surely among the first schemes by which workers were enabled to share in the success of their labours. And at a time when many employers kept their men at work from early morning until late evening its workers enjoyed an eight-hour day.

Union requests for a closed shop were too advanced even for such employers in the 1890's, but the management seems at least to have smiled on the trade unions, if one is to judge from the fact that articles on the various unions represented in Thames Iron Works shops appeared in the TIW Gazette. Drama club, orchestral band, and similar works organisations flourished, but all came to a sudden and catastrophic end just before Christmas 1912, when the hundreds of workers were stood off and the firm went out of business—although it is said that West Ham's first-division football team is the lineal descendant of the former works team.

There were other isolated instances of lifeboats being built by Essex yards, one other Thames yard that played a significant part in lifeboat development being that of R. & H. Green at Blackwall, whose steam lifeboats have been mentioned in the chapter on Harwich.

The first Harwich lifeboat, the *Braybrooke*, was built in 1821 at the Harwich Navy Yard, situated on the tip of the Harwich peninsula. This famous yard turned out warships for the Royal Navy in Pepys' day, and in the 19th century was turning out merchant ships, including steam packets for the Post Office. Probably the building of that early lifeboat caused quite a stir at the time, but it was so far forgotten that when the firm of J. & H. Cann—then building barges and bawleys at the head of Gashouse Creek—built the Frinton volunteer lifeboat in 1907 she was heralded as the first lifeboat ever built at Harwich.

The Navy Yard is now no more, the site being occupied by a modern wharf known as Navyard Wharf, used by bow-loading vehicle ferries, but there are several reminders of the old shipyard to be seen. The wharf proprietors have erected a belfry for its old bell, used to summon the workers in the days when it was a government concern, and its old treadwheel crane now stands forlornly on Harwich Green, to which it was moved when the yard became derelict in the 1930's. The yard returned to life for a few years during the second World War, when some 300 vessels, minesweepers, motor torpedo boats and other naval craft, were slipped for repairs, but the site has now disappeared for ever under the concrete quays. Cann's slip still remains derelict at the head of Gashouse Creek, but it has long been disused.

There remains, however, one yard in Essex which is still concerned with lifeboat work. This is at Rowhedge, on the Colne about two miles below Colchester on the opposite bank to Wivenhoe. The first lifeboat built in the village was the Walton volunteer lifeboat *True-to-the-Core*, constructed at John Houston's Donyland Shipyard to replace the old Walton volunteer boat.

Houston's yard later became part of Rowhedge Ironworks, which built ships for over 60 years after being incorporated as the Rowhedge Ironworks Co. Ltd. in 1905. One of their specialities was the construction of foreign service vessels, river steamers and the like, and it was not until 1939 that another lifeboat was built at Rowhedge. This was a 35 ft. 6 in. self-righter for Cadgwith, the only one of her particular type to be built. In May 1940, when news came of the plight of the British Expeditionary Force, which was being driven towards the sea by the advancing German armies, she was almost ready for delivery. With a crew of Walton lifeboatmen she was taken to Sheerness, from where she was ordered to Ramsgate. Just what work she did during the evacuation from Dunkirk is not recorded, but Charles Vince in *Storm on the Waters* tells how she was towed across the Channel, arriving at dusk, only to be immobilised when the towrope parted and became entangled in the propeller. All night the lifeboat drifted off Dunkirk, her crew watching the fires in the town and listening to the explosions of oil tanks and ammunition dumps. Next morning one of them went overboard with a knife, but he could not clear the rope from the screw; it had been drawn round the propeller shaft and was hard as iron. So the crew made sail, using the lifeboat's own mizzen and blankets sewn together with string and hoisted on boat hooks. Later, as they crept homeward at two knots, the crew managed to loosen the rope a little by working the engine and they were thus enabled to move under power —stern first. In the end the crew were found asleep on a lightship in the mouth of the Thames.

The boat was in due course returned to Rowhedge for repairs, her varnished engine cover riddled with bullet-holes, to the despair of the shipwrights who had lavished such care on her building. Subscribed for by the Girl Guides of Britain, she was to have been named the *Guide*; but when she entered service it was as the R.N.L.B. *Guide of Dunkirk*.

During the war the yard was busy on Government work, but about 1943 work began on the construction of another lifeboat, a 46 ft. Watson type. The decks and beams were in and the forward cockpit was almost complete—she was to have been the first Watson boat with a forward cockpit—when the decision was made to complete her as an orthodox 46 ft. boat without a forward cockpit. She was not completed until 1946, when she was sent to Holy Island.

This boat, the *Gertrude*, was followed in 1949 by the *Southern Africa*, a 51 ft. Barnett type boat which had the distinction of being the first of the Barnetts to be fitted with a diesel engine, earlier craft having had petrol motors. She served at Dover until 1967, when she was replaced by one of the steel 44 ft. lifeboats.

The last of the Rowhedge-built lifeboats was the *B.H.M.H.*, a 35 ft. 6 in. boat of the Liverpool type, which went to Minehead, Somerset, in 1951. Though none have been built at Rowhedge since, there is almost always a lifeboat on the slip or in the shed at Rowhedge Lower Yard undergoing refit.

The WRECK AND RESCUE Series [General Editor: Grahame Farr], are standard histories of Britain's lifeboats, recording the story of each station from its inception to the present day and complete with Boat and Service Records.

Already published —

WRECK AND RESCUE ROUND THE CORNISH COAST

Cyril Noall and Grahame Farr

I. THE STORY OF THE NORTH COAST LIFEBOATS
Bude — Port Isaac — Padstow — Newquay — Hayle
126 pages . 14 plates . maps . 22s6d

II. THE STORY OF THE LAND'S END LIFEBOATS
St. Ives — Sennen — Isles of Scilly — Penzance — Newlyn — Penlee Porthleven
151 pages . 29 plates . maps . 28s

III. THE STORY OF THE SOUTH COAST LIFEBOATS
Mullion — The Lizard — Cadgwith — Coverack — Porthoustock — Falmouth — Portloe — Mevagissey — Polkerris — Fowey — Looe
195 pages . 29 plates . maps . 30s

WRECK AND RESCUE IN THE BRISTOL CHANNEL

Grahame Farr

I. THE STORY OF THE ENGLISH LIFEBOATS
Clovelly — Appledore — Northam Burrows — Braunton Burrows — Morte Bay — Ilfracombe — Lynmouth — Minehead — Watchet — Burnham — Weston-super-Mare
183 pages . 26 plates . maps . 30s

II. THE STORY OF THE WELSH LIFEBOATS
Penarth — Barry Dock — Atlantic College — Porthcawl — Port Talbot — Swansea — The Mumbles — Port Eynon — Llanelli — Burry Port — Pembrey — Ferryside — Tenby — Angle
190 pages . 28 plates . maps . 30s

WRECK AND RESCUE ON THE ESSEX COAST

Robert Malster

THE STORY OF THE ESSEX LIFEBOATS
Harwich — Walton — Frinton — Clacton — Southend-on-Sea — the salvaging smacks — the lifeboat builders of Essex
168 pages . 20 plates . map . 30s

In preparation

WRECK AND RESCUE ON THE COAST OF DEVON

Grahame Farr

THE STORY OF THE SOUTH DEVON LIFEBOATS
Plymouth — Yealm River — Hope Cove — Salcombe — Dartmouth — Brixham/Torbay — Torquay — Teignmouth — Exmouth — Sidmouth

WRECK AND RESCUE ON THE WELSH COAST

Henry Parry

I. THE LIFEBOATS OF CARDIGAN BAY AND ANGLESEY

Aberdovey — Barmouth — Criccieth/Portmadoc — Pwllheli — Abersoch — Porthdinllaen — Llanaelhaiarn — Llanddwyn — Rhosneigir — Rhoscolyn — Holyhead

II. THE STORY OF THE NORTH WALES LIFEBOATS

Cemlyn — Cemaes — Bull Bay — Moelfre — Penmon — Beaumaris — Llandudno — Llandulas — Abergele — Rhyl — Point of Ayr — Mostyn

WRECK AND RESCUE ON THE DORSET COAST

Grahame Farr

THE STORY OF THE DORSET LIFEBOATS

Lyme Regis — Portland — Weymouth — Kimmeridge — Chapman's Pool — Swanage — Studland — Poole

WRECK AND RESCUE ROUND THE EAST ANGLIAN COAST

Robert Malster

I. THE STORY OF THE SUFFOLK LIFEBOATS

The work of the beachmen — Landguard — Woodbridge Haven — Aldeburgh — Thorpeness — Sizewell — Dunwich — Southwold — Kessingland — Pakefield — Lowestoft — Corton

II. THE STORY OF THE NORFOLK LIFEBOATS

The work of the beachmen — Gorleston — Great Yarmouth — Caister — California — Winterton — Palling — Happisburgh — Bacton — Mundesley — Cromer — Sheringham — Blakeney — Wells — Brancaster — Hunstanton

WRECK AND RESCUE ON THE IRISH COAST

J. de C. Ireland

I. THE STORY OF THE IRISH SEA LIFEBOATS

Greenore — Giles Quay — Black Rock — Clogher Head — Drogheda — Balbriggan — Skerries — Rogerstown — Howth — Poolbeg — Dun Laoghaire — Dalkey — Greystones — Wicklow — Arklow — Courtown — Cahore — Wexford — Rosslare — Carnsore

II. THE LIFEBOATS OF THE SOUTH AND WEST

Kilmore — Fethard — Duncannon — Dunmore East — Tramore — Helvick Head — Ardmore — Youghal — Ballycotton — Queenstown —Kinsale — Courtmacsherry — Baltimore — Derrynane — Valentia — Fenit

III. THE LIFEBOATS OF ULSTER AND CONNAUGHT

Galway Bay — Westport — Killybegs — Aranmore — Culdaff — Greencastle — Portrush — Carrickfergus — Groomsport — Donaghadee — Ballywalter — Cloughey — Killough — Rossglass — Tyrella — Newcastle

D. BRADFORD BARTON LTD

FRANCES STREET · TRURO · CORNWALL